Governing for the Environment

Global Issues Series

General Editor: **Jim Whitman**

This exciting new series encompasses three principal themes: the interaction of human and natural systems; cooperation and conflict; and the enactment of values. The series as a whole places an emphasis on the examination of complex systems and causal relations in political decision-making; problems of knowledge; authority, control and accountability in issues of scale; and the reconciliation of conflicting values and competing claims. Throughout the series the concentration is on an integration of existing disciplines towards the clarification of political possibility as well as impending crises.

Titles include:

Brendan Gleeson and Nicholas Low (*editors*)
GOVERNING FOR THE ENVIRONMENT
Global Problems, Ethics and Democracy

W. Andy Knight
A CHANGING UNITED NATIONS
Multilateral Evolution and the Quest for Global Governance

W. Andy Knight
ADAPTING THE UNITED NATIONS TO A POSTMODERN ERA
Lessons Learned

Graham S. Pearson
THE UNSCOM SAGA
Chemical and Biological Weapons Non-Proliferation

Andrew T. Price-Smith (*editor*)
PLAGUES AND POLITICS
Infectious Disease and International Policy

Michael Pugh (*editor*)
REGENERATION OF WAR-TORN SOCIETIES

Global Issues Series
Series Standing Order ISBN 0–333–79483–4
(*outside North America only*)

You can receive future titles in this series as they are published by placing a standing order. Please contact your bookseller or, in case of difficulty, write to us at the address below with your name and address, the title of the series and the ISBN quoted above.

Customer Services Department, Macmillan Distribution Ltd, Houndmills, Basingstoke, Hampshire RG21 6XS, England

Governing for the Environment

Global Problems, Ethics and Democracy

Edited by

Brendan Gleeson
Deputy Director
Urban Frontiers Program
University of Western Sydney
Australia

and

Nicholas Low
Associate Professor in Environmental Planning
University of Melbourne
Australia

First published 2001 by
PALGRAVE
Houndmills, Basingstoke, Hampshire RG21 6XS and
175 Fifth Avenue, New York, N.Y. 10010
Companies and representatives throughout the world

PALGRAVE is the new global academic imprint of
St. Martin's Press LLC Scholarly and Reference Division and
Palgrave Publishers Ltd (formerly Macmillan Press Ltd).

ISBN 0–333–79372–2

This book is printed on paper suitable for recycling and
made from fully managed and sustained forest sources.

A catalogue record for this book is available
from the British Library.

Library of Congress Cataloging-in-Publication Data
Governing for the environment : global problems, ethics, and democracy /
edited by Brendan Gleeson and Nicholas Low.
    p. cm.
Includes bibliographical references and index.
ISBN 0–333–79372–2
  1. Environmental ethics. 2. Environmental justice. I. Gleeson, Brendan,
  1964– II. Low, Nicholas.
GE42 .G68 2000
179'.1—dc21
                                                     00–033359

10   9   8   7   6   5   4   3   2   1
10   09 08 07 06 05 04 03 02 01

Printed and bound in Great Britain by
Antony Rowe Ltd, Chippenham, Wiltshire

For our children
Julian, Ciannait, Vinca, Jennifer

Contents

List of Figures ix

Notes on the Contributors x

List of Abbreviations xiii

1 The Challenge of Ethical Environmental Governance
  Nicholas Low and Brendan Gleeson 1

Part I Environmental Issues, Ethical Dilemmas 27

2 Towards Sustainability
  Joachim Spangenberg 29

3 The Intergovernmental Panel on Climate Change:
  Beyond Monitoring?
  Elizabeth Edmondson 44

4 The International Politics of Declining Forests
  Minna Jokela 61

5 Maximizing Justice for Environmental Refugees:
  a Transnational Institution on Behalf of
  the Deterritorialized
  Adrianna Semmens 72

6 Environmental Accountability and Transnational
  Corporations
  David Humphreys 88

Part II Towards a Global Ethics 103

7 Towards An Environmentalist Grand Narrative
  Arran Gare 105

8 Human Rights and the Environment: Redefining
  Fundamental Principles?
  Klaus Bosselmann 118

 9   Planetary Citizenship: the Definition and Defence
     of an Ideal
     Janna Thompson   135

10   An Ecological Ethics for the Present:
     Three Approaches to the Central Question
     James Tully   147

11   Environmental Ethics and the Obsolescence of
     Existing Political Institutions
     Peter Laslett   165

Part III   Humane Government for the Environment   **181**

12   Environmental Justice and Global Democracy
     Wouter Achterberg   183

13   The Politics of Cosmopolitical Democracy
     Daniele Archibugi   196

14   An International Court of the Environment
     Amedeo Postiglione   211

15   Humane Governance and the Environment:
     Overcoming Neo-Liberalism
     Richard Falk   221

Bibliography   237

Index   251

List of Figures

2.1 Sustainability is a matter of justice   30
2.2 Characteristics of paid labour   35
2.3 From Economy to REconomy and DEconomy – material
    flows in a sustainable society   38
2.4 Taking account of interlinkages is crucial for
    sustainability policies   41
3.1 Organizational structure of the IPCC   47
3.2 Working Group research areas   48

Notes on the Contributors

Wouter Achterberg is Senior Lecturer in Ethics, Political and Environmental Philosophy at the University of Amsterdam. He also holds the Chair of Humanistic Philosophy at Wageningen University (the Netherlands).

Daniele Archibugi is a Director at the Italian National Research Council in Rome. He has worked at the Universities of Sussex, Roskilde, Naples, Cambridge and Rome and he is a consultant of various international organizations including the European Union and the United Nations. Among his recent books, he has co-edited *Cosmopolitan Democracy. An Agenda for a New World Order* and *Re-imagining Political Community. Studies in Cosmopolitan Democracy.*

Klaus Bosselmann is Associate Professor of Law and Director of the New Zealand Centre for Environmental Law at the University of Auckland. Previously, he was a judge, legal practitioner and law professor in Berlin and founding director of the Institute for Environmental Law in Bremen. His recent books include *Im Namen der Natur* (1995), *When Two Worlds Collide: Ecology and Society* (1995), *Ökologische Grundrechte* (1998) and *Environmental Justice and Market Mechanisms* (co-edited with Benjamin Richardson, 1999).

Elizabeth Edmondson lectures in Politics at Monash University, Australia. Her research interests lie within international relations, including the emergence of international environmental regimes, the transformation of states, and international governance mechanisms.

Richard Falk is Albert G. Milbank Professor of International Law and Practice at Princeton University, USA, where he has been a member of the faculty since 1961. He is the author of several books, including *Law in an Emerging Global Village: a Post-Westphalian Perspective* (1998) and *Predatory Globalization: a Critique* (1999). He has been associated with the World Order Models Project since its founding in 1967.

Arran Gare is a Senior Lecturer in Philosophy and Cultural Inquiry at Swinburne University, Australia, and Director of the Joseph Needham

Centre for Complex Processes Research. With Robert Young, he co-edited *Environmental Philosophy*, published in 1983, and since then has published four books on environmental themes, including *Postmodernism and the Environmental Crisis*, published in 1995, and *Nihilism Inc.: Environmental Destruction and the Metaphysics of Sustainability*, published in 1996. His main research interests are environmental philosophy, philosophy of culture and the metaphysical foundations of the sciences.

Brendan Gleeson is Senior Research Fellow in the Urban Frontiers Program, University of Western Sydney, Australia. He is co-author (with Nicholas Low) of *Justice, Society and Nature: an Exploration of Political Ecology and Australian Urban Planning: New Challenges, New Agendas*.

David Humphreys is Lecturer in Environmental Policy at the Open University, UK. He is the author of *Forest Politics: The Evolution of International Cooperation* (London, Earthscan, 1996) and of several journal articles and book chapters on environmental politics.

Minna Jokela is Assistant Professor of International Relations at the University of Turku, Finland. She is completing her PhD on the forest politics of the European Union in a global context, employing constructivist approaches to the study of European integration and global politics.

Peter Laslett is a Fellow of Trinity College Cambridge, and was reader in the politics and history of social structures at Cambridge University from 1966 to 1983. In the 1960s, with others, he developed plans for Britain's Open University, and in the 1970s the University of The Third Age. He has edited five volumes of *Philosophy, Politics and Society: A Collection* (with W. G. Runciman, Quentin Skinner and James Fishkin), and the *Two Treatises of Government* by John Locke, and he has written a number of books on the history of family life. His recent works include *A Fresh Map of Life* (second edition 1996 with James Fishkin), *Justice Between Age Groups and Generations* (with James Fishkin) and *Ageing In The Past* (with David Kertzer).

Nicholas Low is Associate Professor in environmental planning at the University of Melbourne, Australia. He has published widely in the fields of urban planning and ecological politics. His recent book, with Brendan Gleeson, *Justice, Society and Nature* (1998) won the 1998 Harold and Margaret Sprout Award of the International Studies Association of the USA for the best book on ecological politics. Nicholas Low convened the conference on environmental justice of the University of Melbourne in 1997. He has edited two volumes: *Global Ethics and Environment* (2000), and *Consuming Cities* (2000).

Amedeo Postiglione is a Judge of the Italian Supreme Court; Director of the International Court of the Environment Foundation (ICEF); co-ordinator of the Supreme Court's working group 'Ecology and Territory', and Professor of Environmental Law at the Sapienza University of Rome and at the University of Urbino, and Director of the International School of the Environment.

Adrianna Semmens has a background in social and environmental grassroots activism that has focused on community development in diverse, 'diasporic' communities. She has followed this up with doctoral work at Griffith University, Brisbane, Australia. Semmens's research interests include feminist engagements with international environmental politics.

Joachim Spangenberg is by education a biologist and environmental scientist. He has focused on generating and transforming sustainability research results into valid input for the political decision-making process. This includes scenario development and modelling, deriving sets of sustainability indicators and policy impact analysis. Recent research has centred on institutional sustainability indicators, environmental household consumption indicators and future scenarios and models for Germany and Europe with a special focus on labour effects of sustainability strategies.

Janna Thompson is a Senior Lecturer in Philosophy at La Trobe University in Melbourne, Australia. She is the author of *Justice and World Order* (1992) and *Knowledge and Discourse: Defence of a Collectivist Ethics* (1998).

James Tully is Professor and Chair of Political Science at the University of Victoria, British Columbia, Canada. He has published and lectured widely on contemporary political philosophy and its history. A recent work is *Strange Multiplicity: Constitutionalism in an Age of Diversity*.

List of Abbreviations

BCSD	Business Charter for Sustainable Development
CDU	Christlich-Demokratische Union (Christian Democratic Union)
CFCs	Chlorofluorocarbons
DDC	Data Distribution Centre (of IPCC)
FAO	Food and Agriculture Organization (FAO)
FCCC	Framework Convention on Climate Change
FDP	Freie Demokratische Partei (Free Democratic Party)
GATT	General Agreement on Trade and Tariffs
GAW	Global Atmospheric Watch
GEF	Global Environment Facility
GEMS	Global Environmental Monitoring System
GCOS	Global Climate Observing System
GOOS	Global Ocean Observing System
GRD	Global Resource Dividend
GTOS	Global Terrestrial Observing System
HDP	Human Dimensions Programme
ICC	International Chamber of Commerce
ICEF	International Court of the Environment Foundation
IGBP	International Geosphere Biosphere Programme
IMF	International Monetary Fund
IFF	Intergovernmental Forum on Forests
IPCC	Intergovernmental Panel on Climate Change
IPF	Intergovernmental Panel on Forests
IUCN	International Union for Conservation of Nature
MAI	Multilateral Agreement on Investment
NAFTA	North American Free Trade Agreement
NATO	North Atlantic Treaty Organization
NGO	Non-governmental Organization
OECD	Organisation for Economic Co-operation and Development
TNCs	Transnational Corporations
TRIPs	Trade Related Aspects of Intellectual Property Rights
UNCED	United Nations Conference on Environment and Development

UNCTC	United Nations Centre on Transnational Corporations
UNDP	United Nations Development Program
UNECE	United Nations Economic Commission for Europe
UNEP	United Nations Environment Program
WCED	World Commission on Environment and Development
WCRP	World Climate Research Programme
WMO	World Meteorological Organization
WTO	World Trade Organization
WWF	World Wide Fund for Nature
WWW	World Weather Watch

1
The Challenge of Ethical Environmental Governance

Nicholas Low and Brendan Gleeson

Introduction

The last years of the twentieth century witnessed the proliferation of institutions of transnational governance in the politics of the environment, both territorially specific and globally negotiated regimes. Yet, as is noted in the 1999 UN Human Development Report (UN, 1999: Box 5.1, p. 98), global governance in total seems to have gone backwards from 'the architecture of international governance' set up after the Second World War, hence 'reinventing global governance is not an option – it is an imperative for the 21st century' (ibid. p. 97). In this new century global governance for the environment will become both increasingly urgent and increasingly contested. The international (as opposed to global) order of the late twentieth century which provides the basis for negotiation between and among states is in question. Today it is necessary to think both about *global* institutions of governance and about their ethical basis.

The question of global governance has come to prominence in the context of increasing economic globalization, worsening 'macro scale' ecological problems, and new pressures on existing regulatory spheres from migration, wars and natural disasters. The idea of 'global governance', though not new in itself, has been given new urgency by the emergence of problems which transcend the power of the nation state. Among the most prominent are those discussed in this volume:

- the pressing need to move closer towards a steady state economy by the end of the century, and in the short term (the next ten years) greatly increase the resource efficiency of the present economy and the social justice of distribution of its benefits and burdens,

- the need to act effectively to diminish global warming,
- the creation of a regime for protection of the world's forests,
- action on the displacement of population as a result of environmental and other crises,
- the development of compulsory codes of behaviour for transnational corporations.

It is recognized that the state's fading autonomy dilutes the most significant existing source of publicly accountable governance and instates the need for a strong normative dimension in the organization of world affairs. This dilution of government authority is reinforced by the inexorable one-sided globalization of economic power and the weakness and failure of international institutions. Paradoxically it is this very weakness, combined with the one-sidedness of 'globalisation from above' (Falk, 1995), which is causing the 'globophobia', the anxiety about globalization, gripping the consciousness of peoples around the world today.

However, it is not always recognized in contemporary debates on global governance that the creation of new institutions must be accompanied by the further development of human values. The institutional dimension is inseparable from the ethical. At the same time the necessary and crucially important diversity of cultures, the inevitable and beneficial diversity of value systems springing from those cultures, and the acute awareness of the totalitarianism of the 'singular grand narrative', all lead increasingly to demands for an ethical, discursive politics grounded in civil society. This politics in turn requires an ethical 'polyphonic' narrative of a new level and the institutions which will allow such a politics to flourish.

This collection of essays explores one of the dimensions of the value-knowledge system needed in any movement towards humane governance for the planet: the ecological sustainability and integrity of the Earth's environment. The problem of ecological governance at the global level has been addressed by a range of authors and in various international policy fora (e.g. 1992 and 1997 Earth Summits), but the ethical bases for these new arrangements have not been discussed in detail. The discourse on environmental ethics has tended to avoid the issue of global governance, whilst ethical issues have only been weakly developed in debates on international environmental regulation. New ecological values have been advanced, but how are these to be reconciled with the familiar ethical forms – justice, liberty, desert – that have guided institutional practice in the past? Perhaps established ethical formulations and institutional structures resist the possibility of reconciliation. Are new ethical formulations needed for the government of the environment? This book begins from the premise that whilst environmental knowledge and values have developed rapidly, their development must not overwhelm consideration of other core 'humane' values: peace, social justice, and human rights.

This book has three parts. In Part I the authors address a few of the problems which the world will have to confront in the twenty-first century, environmental problems at the national and global scales which pose new ethical issues for governance. In Part II we turn from the problematic to a wider consideration of the ethical basis of global governance and the dilemmas encountered. In Part III potential new global institutions are examined which could embody values of social justice and ecological sustainability.

Environmental issues, ethical dilemmas

Joachim Spangenberg (Chapter 2) sets forth the dimensions of social and institutional change required to fulfil the paradigm of 'sustainable development' during this century. He first points out that not only is the paradigm itself predicated on conceptions of distributional justice (as *between* people now living and those not yet born, and *among* people now living – most notably between the rich 'developed' 'North' and the poorer 'developing' 'South'[1]), but bringing it into being will inevitably cause such social conflicts that distributional justice will need to be constantly demonstrated and contested at every step in the process of change. This will be necessary for each step eventually to be accepted by the key political actors in both North and South.

The world is at present obsessed with the allocation problem and has either ignored the distribution problem or conflated distribution with allocation. Herman Daly correctly points out that efficient allocation is about making sure that what is produced matches what is demanded on the basis of a given distribution of the capacity to pay – income or wealth (Daly, 1996: 159–60). The distribution of that income or wealth is a quite separate issue. Whereas a given allocation is efficient or inefficient, a given distribution is just or unjust. Whereas a market can in principle deliver efficient allocation, it cannot deliver just distribution. There is no alternative to political struggle, deliberation and action to arrive at a just distribution. Just as the market is needed for allocative efficiency, so some form of democratic politics is needed for just distribution. We have to hope too that the third and most critical problem Daly discusses, the scale of economic exploitation of the environment, will also yield to rational and ethical thought via democratic politics because no other tool is effective, safe and available. Certainly the market will not deliver sustainable exploitation. To solve the problems of distribution and scale the world will need all the political and ethical creativity and skill it can muster to develop new political methods.

The scale of required change is formidable: in the European sphere, energy consumption needs to be reduced by a factor of about four, material input to production by a factor of ten ('dematerialization'). But these factors are enormously compounded by current levels of economic growth measured by GDP. Thus a modest annual growth rate of two per cent raises the necessary factor of dematerialization over fifty years to a factor of twenty seven. In truth as Herman Daly and John Cobb (1994: 425–6) point out, exponential growth in the current form is impossible to sustain *under any circumstances*, and will lead to social conflicts as it meets its limits.

Nevertheless Spangenberg argues that ways can be found to make the necessary transformation of the world's economy without revolutionary change of the system of production. The transformation does, however, entail a change in how we conceive of core social values and how they are provided for across every industrial sector, between both the owners/managers of businesses and their workforce, and in what business is done. The focus will have to shift from possessions and their production to the needs and desires that possessions are meant to serve. Labour will change towards increased work sharing and revaluation of time over possession. The economy needs to be refocused. However, 'the reduction of resource use must go with more equitable distribution patterns – or it will go with increasing poverty and thus will be bound to fail, in ethical as well as political terms' (p. 33).

The existing economy has an inertia which can only be countered by strong and sophisticated action by those sections of the political sphere (both inside and outside both state and corporate spheres) which are not themselves subject to the same inertia. This is of course precisely why organizations like the Wuppertal Institute (see Chapter 2) are so important in offering to national states a picture of the future in which capital is compatible with sustainability, so that the power of capital can itself be co-opted to the achievement of this future rather than pulling against it. However movement against the direction of inertia is bound to meet with short-term failures of all sorts. In this respect, an important insight that Spangenberg offers is that failure must be seen not as defeat but as the unavoidable condition of learning. Echoing von Weizsäcker, he argues that life should be open for trial and error and that societal infrastructure must be error-friendly, mistake provoking and forgiving, in order to encourage learning processes'.

The set of institutions which, with international agreement, have been formed around the problem of climate change is truly a global regime of governance, as Edmondson describes in Chapter 3. Unlike the closed and secretive network of organizations governing the world economy (International Monetary Fund, World Bank, GATT, World Trade Organization), the Intergovernmental Panel on Climate Change (IPCC), which forms the 'central pillar' in the creation of the climate change regime, is an open and inclusive organization. Formed in 1988, its membership consists of 190 states and has been strongly supported by the United Nations General Assembly since its inception. Considering the scope and complexity of the scientific, ethical and political problematic of climate change, the IPCC has made a considerable impact on the community of states in a rather short time.

Edmondson argues that particular characteristics of the IPCC play an important role in its success, and by implication in the success of any global institution. First, the working group structure of the organization has been able to change to reflect new perceptions of the climate issues. Second, the organization has been able to build politically relevant scientific knowledge through processes of communication and consultation between scientists and political actors, non-government and state actors. Expert peer review in the scientific tradition is coupled with plural scientific methodologies and political identification of needs by different states, forming a 'marriage between scientific knowledge and policy-making via cross-sectoral activities and communication' (p. 52). Third, this 'marriage' gives rise to stable negotiating coalitions and a capacity to build trust among the participants that the information provided by experts is not merely politically motivated, even if not politically neutral. In these circumstances, reliable knowledge with a claim to objectivity becomes the foundation for the truth, explanation and policy. Finally, the incremental targets agreed in the Framework Convention on Climate Change provide a way of gradual adjustment of national policy, regulation and incentive systems.

Up to now the principle task of the IPCC has been to find out what is happening to the world's climate, why it is happening and what should be done about it, a process which naturally moves towards negotiated actions on the part of nation states to reduce the impact of human activity on the climate. The next step, however, is to move towards verification of compliance of nation states with the agreements they have made, and whether instruments such as 'emissions trading' are in fact working. While Edmondson is hopeful that recent developments in the IPCC are heading in this direction, it remains to be seen whether such a negotiated regime can also bring about compliance, raising questions of target *enforcement*. The chapter provokes a number of questions for environmental governance. Faced with the scientific demand for real and effective action, will the trusting relationship hold between the political and scientific communities that has been built within the IPCC? Or is implementation simply stretching the capacity of a negotiated regime too far? If so what complementary institutions are needed and what role in implementation will the IPCC play? Perhaps the issue of climate change will turn out to be one whose solution demands a shift from *governance* consisting only of loose voluntary networks to *government* involving an institutional structure of authority and accountability (see Young, 1994: 51–2).

Even though forest conservation is as pressing a problem as climate change, the movement towards global governance has been much slower. Perhaps this is because trees are a very concrete 'resource' which can be owned, and are currently being exploited by very powerful business interests, while 'climate' belongs to no one and is self-evidently a global commons. A comparison is instructive because in many ways the forest issue seems to have many of the same scientific-political ingredients as the climate issue – except for the geographical certainty about ownership of benefits and costs (and in this respect see Young, 1999). Jokela, in Chapter 4, describes the halting progress towards a global convention on forests. State interests in Brazil, Malaysia and Indonesia initially blocked progress even towards the creation of an Intergovernmental Panel on forests, framing the issue as a North–South confrontation. From 1995, however, with a new focus on sustainable *use* of forests, the Intergovernmental Panel on Forests was established to foster dialogue. Again, however, there was little progress towards a Convention with the purpose of conservation, but the dialogue continues in the Intergovernmental Forum on Forests. Jokela argues that a lack of clear leadership within the forum is responsible for the lack of progress: 'too many cooks spoil the broth' (p. 65).

The European Union, Jokela believes, ought to play a key leadership role in developing a global forest regime with instruments equivalent to those of the Climate Convention. The Europe of national rivalry, she argues, was the cradle of modernity. Of course the dominant economic theatre later shifted from Europe to the United States, but today, with Union, Europe is poised to play a powerful economic role once more. The very structure and purpose of European Union fits this body to take on the integrative role. First however the Union must itself demonstrate that such an integrated policy, transgressing the interests of states, is possible. Such a challenge is posed in particular by the widening of the Union to include the heavily forested Nordic countries and the still prospective membership of Eastern European countries.

The number of 'official' refugees worldwide has risen on average by 12 per cent per year since the mid-1970s (UNHCR, 1995). It is timely, then, to consider the connection between environmental degradation and the displacement of people. Indeed, as the effects of climate change appear in local environmental destruction, the 'environmental refugee' will become an increasingly familiar category in this century. As Susan George (1977) has observed, 'Hunger is not a scourge but a scandal.' Semmens in Chapter 5 probes this scandal and all its concrete ramifications in a highly personalized and situated way. First, she says, the question of *environmental* refugees is itself homeless in the academic world, hidden in the discourse of 'political' (acceptable) and 'economic' (unacceptable) refugees, and shackled to neo-Malthusian, geopolitical security debates.

Semmens further explores the dimensions of what it means to be a refugee through the multiple stories of poverty linked to displacement. We hear the voices of people who experience this situation not only as victim, but also as oppressor in the immediate sense. But Semmens wants to point out that both are victims of deterritorialization which has two distinct but related meanings. One is that of displacement: the forced migration from kin and social support in place. The other is the disembedded economy without place, whose executives, like the capital they govern, move freely over the globe. It is this broader phenomenon of deterritorialization that appears to Semmens to underlie the problem of environmental refugees. In one perspective, justice might be coping adequately with the symptoms, creating a new UN regime, shaping new rules of 'assistance' – minimal justice according to Semmens. Maximizing justice would address the problem at source in the maldevelopment and impairment of human rights in which the necessity to flee is inflicted. As Amartya Sen has remarked, no famine has ever yet occurred in a democracy with a free press (in O'Rourke, 1996: 81). Unfortunately, though, democracies have all too frequently unwittingly inflicted famine elsewhere.

Addressing the global economy more directly, Humphreys (Chapter 6) notes the growth of the power of the transnational corporations under the regime of neoliberalism following what Lasch (1995: 45) terms the 'revolt of the elites', or the 'Second Glorious Revolution' (van der Pijl,

1995). Humphreys considers various contemporary models of regulation of transnational corporations: codes drafted by corporations themselves (such as the Business Charter for Sustainable Development), codes drafted by civil society actors (such as the proposed Treaty on Transnational Corporations), codes drafted by corporations with civil society actors (the code of the Coalition for Environmentally Responsible Economies), and the attempts at drafting a code of conduct for TNCs within the United Nations. Humphreys finds all of these attempts, for different reasons, inoperative or ineffective.

Arguing against the view that states are powerless in the face of 'globalization', Humphreys suggests that the national state must again play the key role in regulating TNCs for purposes of sustainable development conceived in terms of environmental conservation and social justice. A strategy of 'redomestication' should be pursued in which corporations should be required to abide by a 'public charter' in order to operate, in effect asserting the power of the state to license the operations of TNCs in quite a broad sense. The severance of the currently close relationship between business and party politics is a necessary condition of effective legislation on the part of the state.

However, reasserting the state role in regulation is necessary but not sufficient to construct global governance for ecological sustainability. Both environmental problems and TNC activities extend over state boundaries. The state regulatory role itself is jeopardized where there are differences in environmental standards among states. It is in the short-term interests of corporations to escape high environmental standards and they will be reluctant to adapt their behaviour to comply with such standards unless their rivals are also subject to the same rules. Thus a strengthening of international law is also required. A move towards cosmopolitan democracy would assist this strengthening and at the same time begin a process of extending public democratic law beyond state territories (see Chapters 12 and 13). Such a move would seek formally to distance corporations from the shaping of the law that is to govern them, just as a democratic national parliament grants no privileged access to business interests. It would entail increased democratic governance within the corporation, and the creation of cosmopolitan law and jurisprudence via new institutions such as a directly elected chamber of the United Nations and an International Environmental Court.

The idea of an International Court of the Environment is canvassed below in Chapter 14. We have suggested elsewhere that such a court is urgently required (Low and Gleeson, 1998: 191). If it is to have the necessary authority, international environmental law should issue from a democratic forum subject to direct election. We have proposed a World Environment Council to deliberate on behalf of the people of the world, and to formulate and declare environmental law. The transition from governance to government implicit in such a step is however a momentous change at global level which requires searching ethical examination.

Towards a global ethics

Approaching solutions to the dilemmas posed in Part I in a rational way appears to demand a new discourse of modernity in place of the neoliberal discourse which governs the world economy. One such discourse is that of 'sustainable development' and the various instruments and institutions necessary to make it effective. However, as we rush towards solutions, postmodernist philosophers apply the brakes of scepticism. Gare in Chapter 7 starts with the paradox posed by the postmodernists: if modernity is characterized by the striving for control over nature and people, and modernity, with its increased capacity for control, is also *producing* the ecological crisis, then how can we possibly resolve the crisis by adding more control? Can the cause of a problem be its solution?

This paradox leads Gare to examine the function of such discourses or 'grand narratives'. He finds that narratives about life beyond the individual are somehow necessary to the fulfilment of the individual life. The search for meaning in the larger temporal, spatial and social picture into which the individual fits appears to be fundamental to the nature of the human species. So narratives are not, or need not be, merely the search for control, though undeniably meaning and control are intimately connected. Interestingly Gare cites as precursors of European civilization, societies where meaning and control have been most closely connected: 'China, India, Babylon, Egypt, Israel and Greece.'

Unfortunately postmodernism, by denying the validity of any new discourse, or 'grand narrative', with the scope and power to unify the forces of opposition to neoliberalism, simply leaves the field open for its triumph. What then is to be done? Shouldn't we, Gare asks, 'promote a subversive "rhizome" politics eschewing any overall strategy?'. Gare's immediate answer is negative: such tactics have been 'disastrous' and 'almost totally ineffectual'. Yet Gare thinks that postmodern scepticism of modernity is justified. This tension leads Gare to explore the possibility that it is not 'grand narratives' as such, but rather some particular aspect of the grand narratives of modernity that is at fault. He concludes that such is indeed the case and that a new kind of grand narrative – not 'monologic' but 'polyphonic' – is required for the further development of the quest for truth and justice as economy and society as currently constituted confront their environmental limits.

The question of the extension or obsolescence of the human rights discourse characteristic of modernity is the concern of Bosselmann in Chapter 8. Bosselmann explores rights as a foundation for global governance and law. A key distinction here is between the exclusive anthropocentric tradition of *human* rights and ecocentric conceptions of nature (inclusive of humanity). While the idea of 'environmental rights' may be easily reconciled with the human rights tradition – the environment being merely another 'good' for human use – there is a potential conflict between human rights and 'ecological rights'. While, in the former case, human rights may be limited only in the interests of enlarging the freedom of all humans, in the latter case limitations on human freedom are justified for reasons other than human freedom. This is because, in the ecological conception of rights, beings other than human are considered to have intrinsic value and their freedom to develop in their own way must also be considered.

At first sight embracing ecological rights (rights of nature) appears to require a radical paradigm shift, improbable if not impossible without some vast catastrophe engulfing humanity. Yet Bosselmann's exploration of constitutional debates shows that, paradigm shift or not, ecological rights have already become part of the discourse on new national and international constitutional developments. Of course, as he says, the burden of proof is on those advocating ecological rights: 'What is the advantage of ecological human rights? Would they make any difference for the real outcome of decision-making?' (p. 129). Bosselmann argues that individual freedom is conditioned not only by a *social* context (which gives rise to the claim of solidarity and welfare rights) but also by an *ecological* context. The ecological context can no longer be regarded as beyond human harm, a kind of omnipotent and indestructible matrix of individual human endeavour. Recognition of that reality leads to a desire on the part of humans to respect the ecological context just as the social context is now respected – at least in the spirit of constitutional law. A new tradition of ecological respect might grow within the existing social order with growing scientific understanding of the fragility of the natural world. Such a movement would ease the transition from an anthropocentric modernism which exhausts the natural context of the economy to an ecological post-modernity which sustains it, together with its social context, as 'sustainable development' demands.

We live in a world of many cultures and, if the communitarians are at least partly right, from these different cultures different values and loyalties grow. Today we regard such multiplicity as a virtue. Yet if there is a global interest in humanity and its earthly environment, as intuition suggests, then we also have to find values which transcend national cultures. This issue is addressed by Thompson in Chapter 9. The work of Held and Archibugi on cosmopolitan democracy forms the backdrop for both this and the next chapter. Thompson interrogates the tradition of cosmopolitanism. While critics of cosmopolitanism distrust claims to universality of values and the likely beneficence of global governmental institutions, some environmental problems are ineluctably global in scope, many more are transnational, most are rooted in a global economy which has drastically reduced national local autonomy. What seems to be needed, then, is an ethic which bridges between the global and the local. Thompson explores one possible ethical 'bridge', that of 'planetary citizenship'.

Thompson discusses a number of objections to the idea of 'planetary citizenship'. Yet, she argues, people across the world do in fact seek the means of cooperation to put into effect their emergent sense of responsibility for the natural world, their common heritage, and their duty to others both present and future. There is no reason today why cooperation should be limited by national boundaries – and it patently is not. If in fact we seek the means of cooperation, we thereby create a sense of planetary citizenship. It is a *resultant* of action rather than its goal. Thompson writes, 'A planetary citizen is someone who assumes her share of responsibility for the collective achievement of goods which she and virtually everyone else values' (p. 145). Planetary citizenship does not replace membership in national or subnational communities, nor does it imply common identity. Thompson argues that planetary citizenship should be understood as a natural development from relationships and responsibilities which individuals already believe they possess and which cannot be protected and carried out without global cooperation. Sometimes regional or global interests will take precedence over local or national objectives, but people will recognize that some sacrifices are necessary for the sake of the values that they possess as members of a community or a nation. If planetary citizenship is ever realized it is more likely to stem from the actions of citizens than of governments, actions taken by those who recognize the inadequacy of their own efforts within existing institutional frameworks to preserve what is important to them.

In Chapter 10 we turn from the practice of cooperation to that of conflict resolution. James Tully confronts the question: 'how can the fact of cultural and philosophical difference on justice and nature be reconciled with the urgent need to deliver fair judgements in cases of conflict between development and the environment, exploitation and conservation?' Like Thompson in the previous chapter Tully begins by reviewing criticisms of cosmopolitan democracy. He points out that the time horizon of cosmopolitan democracy is long term rather than immediate, and is premised upon the singular value of autonomy. The priority of such a value can be challenged from several points of view. First its singularity can be questioned: don't other contending values have a right to be ranked highly? Second, many ecologists would indeed rank 'interdependence' as a higher value than autonomy. And, third, it can be argued that no value should be above democratic debate, even such an overarching value as 'autonomy' – though here it is fair to ask if the practice of 'democratic debate' might itself be premised on the assumption of autonomy.

However, given that democratic debate is the value to be sought, Tully seeks ways in which 'people's background conceptions of justice and nature' may be brought into the discussion 'and criticized through a reasoned exchange' (p. 149). What is needed is not so much political pluralism – a tacit 'agreement to differ' – as an active engagement between different fundamental conceptions of nature if we are to arrive at practical decisions. Here Tully confronts the arguments of Rawls and Habermas concerning the primacy of individual morality over communal ethics. Ethical reasoning, Tully argues, is necessary in order to resolve upon a common community interest, indeed the global interest of humanity in nature. What is needed is the confrontation in argument between ecocentric and egocentric perspectives, not coexistence. Moreover such a confrontation cannot be conducted outside power relations, since action also requires power, though they may be conducted by means of practices in which domination is minimized. Tully deploys Foucault's strategies for understanding power within 'practical systems', that is organized forms of human activity and relations of communication through which that activity is coordinated.

What then should we do? Tully urges us to start from actual contests over ecologically damaging forms of conduct, accept value plurality and representative democracy – but representative also of non-humans and future populations – allow for the critical discussion of core differences over justice and nature, and in the process come to fair judgements. Required is an experimental and prudential *ethos*, in which 'critical reflection on one experiment in modifying our relation to nature will provide the basis for the next' (p. 152). This is no simple task. In particular it is important not to fall into 'broad oppositions' such as 'development versus environment'. The core principle should always be to listen to the other side, and listen carefully, giving due consideration to legitimate concerns. The practice of reaching a reasonable judgement will not result in a definitive resolution of the central question or a consensus. There will always remain an element of reasonable disagreement, and therefore the possibility of reasonable doubt and dissent. Any judgement will be a negotiated accommodation or reasonable compromise. Tully resists the characterization of global capitalism as a coherent system: in his view this is an inaccurate picture as well as disempowering for local actors. It is primarily our 'routine acting' within practical systems that we need to examine and change.

Addressing our practical systems in a somewhat different way, Peter Laslett in Chapter 11 considers the obsolescence of existing political institutions. Laslett discusses four 'straightforward' propositions. The first is simply that, if we are to deal justly in respect of the relations among humans and between humans and the non-human world, we must have authoritative and effective institutions. Though this proposition may seem self-evident, the contrary opinion advanced by the founders of anarchism deserves careful scrutiny. Though we can hardly await the emergence of an ideal political world, there is scope for much more collaboration on environmental matters outside the state sphere than currently takes place.

Second, our political institutions show signs of losing their power to act in the pursuit of environmental justice. Nation-states and the institutions of the United Nations seem unable to control the activities of transnational corporations, some of which have a resource base greater than many nations (see Chapter 6). However it must also be said both that nation states themselves have committed assaults on our ethical relationship with 'nature', and that some corporations and those who run them have shown considerable willingness to protect the environment. We should not cynically dismiss these efforts. The third proposition is that existing institutions are in some ways inappropriate for the tasks of environmental justice. The 'nation state' encompasses both states of enormous population and power, and micro-states, of population not much bigger than a municipality in the larger states. The founders of modern institutions certainly had in mind that these institutions should endure far into the future, but the concept of a *limited* environment is new, and with it the idea that humans had to consider the justice of a *shared* environment and the justice of their relationship with the non-human world.

The fourth proposition is that environmental matters affect all humans regardless of culture or polity. Boundaries among nations are rendered arbitrary by international and global environmental threats. Environmental threats confront us with the vulnerability we all have in common. The effects of what we do now will last for an extremely long time, perhaps for ever. Existing institutions were not and cannot be designed for eternity and must to some degree be defeated by the timescale of environmental challenges. Moreover, because environmental ends can only be approached by agreements reached between and among governments we have no alternative but to look for solutions to the 'rickety edifice' of international authority.

Laslett is sceptical of both the organizational and the representative capacity of the liberal democratic nation state (as currently constituted), and of any order negotiated among such states, to solve environmental global problems. If the current nationalistic world order is powerless over such problems, it seems as though the only way forward is to sweep away all structures of governance and begin again. Recognizing that such a proposition is hardly constructive, Laslett looks for an answer in emerging forms of political representation and ethical thought. For example, in 'citizens' juries' and sample 'deliberative polls' people could participate – even at a global scale – and communicate with one another not as citizens of nation states but as members of a virtual world human community. The religious mode of thinking is even now beginning to infuse the discourse of environmental ethics. Laslett foreshadows a kind of religion not attuned to domination and aggression but drawing on environmental revelation. Deliberative polling could provide the channel through which the public voice on matters of ecological spirituality could be made known to those with worldly power, a voice it would be hard to ignore.

Humane governance for the environment

In the first eleven chapters we move some way from the discussion of environmental issues towards an exploration of the dilemmas of global ethics. In the final section of the book the authors delineate forms of humane global governance for the environment. The term 'humane governance' comes from the contributor of the final chapter, Richard Falk, whose work on global governance has been well known since the World Order Models Project of the 1960s of which he was a leading author. The idea of humane global governance, as Falk explains, takes nothing for granted, in particular the centralized statelike 'institutional hardware' (Dryzek, 1999: 277) associated with 'government'. There are reasons, in our view, why increased authority at global level may have to be considered. We do not think that the reasoning behind the World Order Models Project, which embraced such a democratic central authority – equipped with many checks and balances – is so very obsolete, although the timescale for its creation was undoubtedly much underestimated.

Achterberg in Chapter 12 argues for a liberal-egalitarian conception of environmental justice and explores the connections between environmental justice and global democracy. His point of departure is the 'egalitarian plateau' which Dworkin (1983: 24) defined as the common basis of all serious modern political theories, the presumption that 'the interests of members of the community matter, and matter equally'. On environmental justice he first compares two principles derived respectively from Luper-Foy and Pogge. The first, from Luper-Foy, states that resources are to be handled in a way that is equitable across the globe and across generations: the 'resource equity principle'. Resources are to be understood both as inputs to economic processes and environmental sinks for waste. The emphasis here is on intergenerational equity. A ceiling is placed on the rate of economic exploitation of the natural world for each generation. The second principle, from Pogge, states that all persons in the world are entitled to have their basic needs met – food, clothing and shelter. To implement such a principle requires a Global Resource Dividend (GRD) funded by a tax on the consumption of natural resources. The GRD would be provided by national governments and would expect to raise some $300 billion per year.

The problem is that the principles of intergenerational and intragenerational justice may point towards different actions and priorities. Reducing consumption of the environment might, for example, make it more difficult to relieve global poverty. How can the two principles be reconciled? Achterberg appeals to the proposition evinced by Henry Shue: while it would be fair to expect poorer countries to pay for their use of the environment in order to protect the interests of future generations, it would not be right to ask these countries to slow down their rate of consumption of the environment to solve ecological problems caused by the past actions of rich industrial countries (global warming for example). At the very least the cost of 'coping' with such problems ought to be borne by the rich nations. Therefore sustainability (inter-generational justice) may be considered a precondition for abating poverty, but only subject to Shue's proviso.

What does such a conception of environmental justice demand by way of governance? And how is environmental justice connected with a democratic order? First, given the variety of arguments that can be made on behalf of environmental justice, and the variety of decisions to which these arguments lead, adequate legitimation of such decisions demands that they be made only after due deliberation and an agreed process leading to a rational consensus among all affected: in short some kind of deliberative democracy. A viable and legitimate concep-tion of environmental justice seems to imply a strong and vibrant deliberative democracy. But does the reverse apply? The Commission on Global Governance has suggested that poverty is likely to under-mine democracy, though the outcome of the historic referendum in East Timor has since shown that this is not always true. Still, environ-mental justice does seem to be a prudential requirement for a stable world order supportive of democracy, though not necessarily *liberal* democracy.

Achterberg examines Held's proposal for 'cosmopolitan democracy' and finds that the GRD proposed by Pogge is in keeping with the spirit of the principle of autonomy on which cosmopolitan democracy is based. Cosmopolitan democracy, providing a common structure of political action, would limit the sovereignty of nation states by addi-tional tiers of democratic governance at regional and global level. Moreover, the departure from liberal democracy towards some democ-ratization of the economy is encouraging from the point of view of environmental justice. More problematic is the transition from present arrangements of global governance, and here Achterberg finds Held's theory somewhat utopian.

Archibugi in Chapter 13 starts by reminding us of the power and authority of the nation state, 'a recognized institution that is the only one authorized to use force' (p. 200): truly a remarkable human inven-tion. The state is where we have to start looking for democracy, yet although democracy has achieved much progress inside states, very lit-tle democracy has been achieved in the international sphere. Archibugi asks why this is so. Is it because of the existence of authoritarian states with which democratic states must deal in an authoritarian manner in order to protect the national interest? This does not appear to be so, since the way democratic states deal with non-democratic states varies not according to the type of government of the latter but according to how well it suits the perceived interests of the former. Why should Serbia, Archibugi asks, be bombed, while Turkey is a member of the alliance doing the bombing? Not obviously on account of the relative democracy of the two countries. Something more than internal democracy within the nation state is called for.

The cosmopolitical project is an attempt to apply some principles of democracy internationally. Archibugi now favours the term 'cosmopolitical' rather than 'cosmopolitan'. The reason is that 'cosmopolitical' places more emphasis on the need to create a democratic international politics rather than comprehensive democratic global constitutions. Cosmopolitics seeks to reformulate the principles on which democracy was founded within states to apply at global level: 'extending democracy globally means designing a form of organization of the political community which, unlike the traditional one, seeks to do more than reproduce the state model on a planetary scale' (p. 204). The project seeks to create institutions which allow the voice of individuals to be heard in global affairs irrespective of the voice they have at home as citizens or subjects.

To achieve cosmopolitical democracy means depriving states of their oligarchic power internationally, enlarging democracy both within states and in interstate relations, and introducing democracy at global level. Considering how transnational economic interests and military power are today globally organized it is surprising, Archibugi observes, that political parties are still almost exclusively a national phenomenon. Forms of political representation therefore remain locked inside state frontiers, and even in the European Parliament parties operate on a national basis. Yet this may change, and probably the more so if popular organizations come to exert real power through global institutions. It is not difficult to imagine that in the coming century what are now non-government organizations might take on some of the character of political parties if their representatives were to have some real power in global forums. But such power is unlikely to be forthcoming unless these forums are democratized.

The evident movement of the past twenty years towards intervention by ad hoc military coalitions in states' affairs on humanitarian grounds indicates a decline in the Westphalian concept of sovereignty. Yet the authority for such interventions is extremely tenuous, and consistency of principle and humane process in deciding when and where force is justified is notably lacking. War against a nation is grossly inhumane when police action against individuals only is warranted. And of course the certainty of consequences following murderous abuses of human rights is altogether absent, so interventions have the character of revenge after the event rather than a final act in reinforcement of a universal deterrent enacted by the certainty of the guilty being brought to justice. What is lacking in cases like the Gulf War and the conflict with Serbia over Kosovo (and now the East Timor intervention) is a total lack of relationship between the *prima facie* culprits of crime and the suffering inflicted. The innocent citizens of 'targeted' states are not protected, increased rather than reduced violence is provoked by the threat of intervention, the guilty are not brought to justice, and the intervention generally results in a materially worse situation than existed before the intervention took place. This shocking situation can hardly be allowed to continue. Therefore proper institutions of justice to deal with massive abuse of human rights – including environmental rights – must surely gradually develop.

Thinking in terms of less comprehensive, more incremental institutional steps that could be taken to institutionalize global environmental protection, Postiglione posits an International Court of the Environment (Chapter 14). As himself a judge of the Italian Supreme Court, Postiglione sees that the International Court at the Hague was not constituted to adjudicate the kind of conflicts over and threats to the environment we face today. The nature of environmental disputes is such that damage done to one part of the environment is frequently damage done to humanity in general. The problem of adapting legal systems, up to now based upon nations, to dealing adequately with questions of global significance and significance for future generations is today being posed for the first time. What is needed is the power to apply international law to the environment with authority and effectiveness by impartial judges who act on behalf of a supranational authority.

The conditions for the creation of such a power are, among other things, a framework convention for environmental law drafted by the states, a supranational court with wide powers of decision, accessibility to individuals and non-government organizations, and a body of non-removable judges. Most of these conditions are lacking from the current constitution of the International Court at the Hague. The International Court is principally a court of arbitration, access is reserved only to states, and its jurisdiction is in any case recognized by fewer than one third of the states of the world. The International Court, for it to become effective in the delivery of ecological justice, would require not just reform but refoundation.

The resolution of environmental disputes would require the use of alternative methods of dispute resolution other than adjudication: for example negotiated settlement and arbitration. However, in national legal orders arbitration and other methods do not *replace* the normal process of adjudication but are complementary to it. Resort to adjudication remains open to the parties. Arbitration is essentially a private and voluntary process, while what is needed for the environment is a transparent public process. The arbitrator's decision is limited only to the parties while the decisions of the proposed International Court of the Environment would have application over all (*erga omnes*) where it has to deal with environmental problems of global scope. The Court would enable the exercise of every human person's right to the environment and would provide a guarantee of public notice and transparency. The Court would create an innovative and evolutionary jurisprudence helping to develop common principles in this new field of law. Finally an International Court of the Environment would perhaps follow the example of the European Court of Justice in establishing the principle of supremacy of international environmental law over the domestic law of the member states. Implicit in the proposal for an International Court of the Environment is also an international forum established on a voluntary basis by states to declare the existing international law on the environment. Such a forum could be established by some states initially as an experimental basis for an Environmental Court.

Effective governance for environmental conservation and even sustainability is not necessarily *humane* governance. It is possible to imagine a situation where consumption is limited to the rich few in the world and an authoritarian regime preserves unequal access to resources in the face of increasing scarcity. Humane governance, in Falk's concept (Chapter 15), gives recognition to the right of all people to life, adequate subsistence and an international order capable of protecting those rights. In order to move in the direction of humane environmental governance it becomes urgent to move away from the neoliberal world order in which a microeconomic rationality is imposed over all aspects of social life. Falk seeks a post-Westphalian world in which states are content to relinquish aspects of their sovereignty, yet are re-empowered to act for the protection of the people and environments within their territory, but also, because the environment is indivisible, beyond it: 'a re-empowered state would act alongside other political actors, including those representing civil society, and have as its most urgent mission the negotiation of a global social contract with market forces that would include environmental protection as a vital element' (p. 223).

The obstacles to such an outcome, Falk points out, are considerable. States have become instruments of the private sector, and in competition with other states, have lost the capacity to promote the public good and its environmental aspects in particular. The United Nations organizations bearing on environmental protection are weak. UN conferences on the environment have raised environmental awareness among states but their influence diminishes over time. Some negotiated regimes have had some success, for example the ozone protection regime, the Law of the Sea and the governance of Antarctica. But the more comprehensive and global cooperative system needed to produce an effective climate regime has not been achieved. Protection of the environment, Falk says, interferes with the workings of the market. Strictly, though, such protection interferes with vested interests operating in the market to prevent appropriate limits being set on its operations, but it is certainly true that neoliberal dogma resists any interference with so-called market forces. Moreover, environmental threats are of long duration and geographically wide scope, and there is little acceptance of the 'global public good' which it is now the duty of states to sustain.

What first steps, Falk asks, might now be taken towards humane governance for the environment? One such step is to reaffirm the 'helpful normative architecture' of the kind provided by the UN conferences. Falk lists the ideas promulgated at the Rio Earth Summit which have informed states' rhetoric but not, so far, much in the way of behaviour. The implementation of normative ideas is dependent on pressure from below by transnational social forces and fears generated by perception of environmental disasters. Much, in Falk's view, now rests on the energy that can be mobilized outside the state sphere to persuade states to give environmental and ecological norms real effect. Adaptive responses which do not challenge neoliberal governance involve privatization and volunteerism: the trading of rights to pollute and appeals to business leaders and the super-rich to act on environmental norms to compensate for state failure. These initiatives, though marginally helpful, are not likely to change the economy in the way which now seems necessary.

What is required is a more fundamental transformation. In this respect neoliberal governance, in which state actors accept a passive role, stands as a major obstacle. Signs of hope, however, are emerging that neoliberalism is beginning to retreat. The world financial and political instability in the wake of the Asian economic crisis of the late 1990s has brought new questioning among business elites about the wisdom of running the economy on 'automatic pilot'. Falk discerns a new internationalism of 'globalization from below' with the formation of coalitions between large groups in civil society and states. The European model of compassionate regionalism offers hope of an assertion of geographical justice. 'There is little doubt', Falk writes, 'that European regionalism is already a far more daring and radical experiment in restructuring world order than anything associated with the United Nations' (p. 235). These fissures in the neoliberal edifices of authority offer hope for a transition to humane governance on a broad front.

The challenge of ethical environmental governance

The discussion pursued in this book raises questions which pose a colossal challenge to humanity in the twenty-first century: the challenge of ethical, environmental governance. First, though, to recapitulate, why are ethics critical to ecological governance? Should we not simply 'get on with the job' of saving nature from humanity or vice versa? Are not the issues self-evident? Marxists might ask: should we not simply root out a political economic system that is self-evidently ruinous? Why delay ourselves with ethical discussions?

We should always keep the ethics of politics under review and subject to open debate because to do otherwise is to risk inhumane governance. Political action in pursuit of apparently self-evident truths was represented most strongly in the twentieth century by the politics of Stalinism, National Socialism and Fascism. To say that such politics has a blemished record of human rights abuse is understatement of the highest order. Even recent humanitarian interventions on the part of democracies have failed to prevent humanitarian disasters. In the creation of institutions with the power to act effectively it is imperative that the ethics both of those institutions and the process of their creation be kept under constant scrutiny. The authors of this volume are in agreement that the first imperative is to create the possibility of a politics in which many voices can be heard, in which dissent is normal, but in which discussion can eventuate in action on the environmental problems the world confronts. Such a 'cosmopolitics' embodies an idea of planetary citizenship in which what humans share, the risks and dangers as well as benefits of use of the environment, becomes of mutual concern.

Inaction on the environment is today not a viable possibility. The environment is already exercising its own imperative. Both climate change and biodiversity depletion have vastly ramifying economic implications and are already provoking disquiet among powerful actors in the world economy (Elias, 1999). Yet action of the kind required brings new perils, and the institutions we create must protect humanity from them. The precautionary principle is as much warranted in efforts to transform human institutions as it is in efforts to exploit the non-human world. Only through the precaution of critical ethical debate will it be possible to create institutions which are themselves sustainable and which develop over a long period in such a way as to protect both human rights and the Earth's environment. Yet in the end, if one is serious about such concepts as environmental rights and citizenship, change of institutions will surely become necessary.

Why do political parties not form outside the nation state? Perhaps because there is no opportunity for such parties to gain real political power outside the nation state. Institutions not only respond to but create politics. Democratic reform of the United Nations, the establishment of an International Court of the Environment, and a World Environment Forum are not only useful starting points for institutional reform but may also empower global politics in civil society. In the longer term global taxation must be seriously considered and with it global representation. Mechanisms such as deliberative polling that enhance learning and employ global sources of communication must be further developed and used in problem solving. All the more important then that access to such communication channels is widely and equally shared. The Internet must not remain the tool of a global elite. The right to free communication should be an early candidate for global protection.

Humanity has developed political ethics (variously framed) for political scales up to the nation state, different of course in different national cultures. But now the globalization of certain aspects of human culture (economic globalization) and an appreciation of the global relationships within non-human nature and between humanity and nature (ecological globalization) have overreached these scales, requiring a new political scale (supranational) for which we must have a political ethical framework. The climate negotiations are a key example of the ethical complexity that global environmental governance reveals. These negotiations also reveal the limits of voluntaristic governance, pointing inexorably in the direction of *government* based on democratic authority and accountability.

Behind the possibility of government in some form at global level, however, there are dialectical oppositions which can only be reconciled in what must always be considered a provisional politics – even of institutions, even of constitutions. There is no such thing as a thousand-year regime, and even a thousand years is just an instant in ecological time. First, then, there is the opposition between change itself and permanence. If an institution is to be effective it must have a certain durability. It must to some degree be insulated from the everyday pressures of the politics of environmental issues. It will not do for a party who is judged at fault under the environmental rules immediately to seek to change the rules. Yet those rules need to be kept under review. It must be possible to change constitutions, yet the politics of constitutional change must proceed at a much slower and more deliberate pace than the politics of issues. Just as there is a spatial dimension of environmental governance so also there is a temporal dimension.

Then there is the opposition between justice in the distribution of environmental values to humans and justice to non-human nature. The injustice of distribution of good and bad environments is not merely an injustice which occurs at the end of a process of production. It is embedded in the process of production itself. We are accustomed to assuming that commodities belong to someone and that the environment belongs to no one or to everyone, but this is only true if 'belonging' is understood exclusively in terms of property rights which confer an absolute instrumental sovereignty over a thing. However the term 'belongs to' can be used *without* invoking property rights. 'Belongs to' in this non-proprietorial sense means being affected by and having a relationship with. As we have argued elsewhere, the environment is constitutive of the self (Low and Gleeson, 1998: 148). In this sense the environment always belongs to someone in some place at some time. So the flow of commodities through an economic process of transformation takes with it elements of people's environments. The gold ring contains a proportion of the environment destroyed to get at the gold (Schmidt-Bleek, 1994). The newspaper contains a proportion of the old growth forests that supplied the woodchips to make the paper. Each litre of petrol consumed contains a portion of the atmosphere used up as a sink. The environment which belongs to people – in specific places and times – in a non-proprietorial sense is being expropriated for nothing and incorporated into commodities which become the property of different people. One class of people, one part of the world, in one time period (property owners, in the 'North', past and present) have benefited enormously from this process at the expense of the environments of another class, in other parts of the world and in another time (non-owners, in the South, and people of the future). There is an environmental injustice at the core of the process of economic growth.

But why should we think that 'the environment' is constitutive only of humans? One of the best-understood dimensions of species extinction is destruction of habitats – which are precisely species-constituting environments. Humans are very good at creating artificial habitats which nourish a very small range of species bred for human consumption. The numbers of people on the planet is no doubt attributable precisely to this skill. Otherwise Malthusian logic would apply and the human population would be kept in balance with that of other predators at the top of food chains. Expropriating the habitats of other species is itself constitutive of humanity or has been up to now. Is this a feature of humanity that must endure or is it one we humans will have to think about giving up or at least modifying? If so, who is going to be the first to do so? If this is what justice to nature demands, does an ethic of justice apply at all outside an anthropocentric perspective? These kinds of question are beginning to be asked. Andrew Dobson (1998: 238), for instance, in his illuminating discussion of the applicability of a justice ethic to ecological sustainability, finds only limited compatibility between justice and an ecocentric environmental ethic which regards all of nature as having intrinsic value.

Arguments over substantive issues of justice are striking out into new fields. These arguments will not resolve upon some universal formula, self-evident once it is discovered. What is important is that free debate should continue and have some input into the political processes by means of which environmental conflicts are resolved. If Marx was right in saying (in the *Theses on Feuerbach*) that 'philosophers have only *interpreted* the world, the point is to change it', *changing* the world also requires philosophical reflection linked to political deliberation. Attention must be paid not only to the substance of justice, justice of outcomes and consequences, but also to the justice of procedure.

In this volume our authors, drawn from a range of countries and professional contexts, contribute to this free debate that must necessarily interpret and guide a changing world. However, the reader will search in vain for simple unanimity as a variety of opinions, theoretical inclinations and policy prescriptions emerge in the chapters that follow. Nonetheless, a common point of departure is evident, each essayist agreeing on the need for ethically-informed institutional solutions for global environmental problems. Moreover, it is agreed that a social ethics, not simply a personal moral outlook, must be harnessed to the task of designing novel forms of environmental regulation, a need not yet entirely recognized in ecological philosophy.

After their separate journeys through distinct, though often overlapping, theoretical and policy landscapes, our authors arrive on broadly common ground, which suggests to us that *government* not governance must be the foundation for a new institutional regulation of nature and its human socialization. The situation of the nation state, and its subsidiary institutions, may have been radically changed by globalization but governments remain the custodians of justice, the defining quality of democracy laid out by the Ancients and in Enlightenment thought. Private and voluntary institutions may act justly – and we urge them to do so – but there can be no justice – and therefore no sustainability – without a vibrant and democratic public sphere.

In the twentieth century it was recognized that a truly democratic *government for the people* was needed at a variety of scales to address human social failings: welfare states were constructed to safeguard material needs and the United Nations established to secure basic human rights and to prevent wars. In the new millennium a new challenge awaits us, to address another dimension of human folly that threatens the entire planet. Enacting *government for the environment* is a task that we cannot delay.

Note

1. 'North' refers to North America, Japan and Western Europe, while 'South' refers to South America, India, sub-Saharan Africa and South East Asia. That there are today major exceptions in Russia and China in the northern hemisphere and Australasia in the southern hemisphere makes it necessary to place 'North' and 'South' in quotation marks.

Part I

Environmental Issues, Ethical Dilemmas

2
Towards Sustainability
Joachim Spangenberg

Introduction

'The Brundtland Commission defined sustainability more than ten years ago as essentially a normative concept' (World Commission on Environment and Development, WCED, 1987). Under this definition 'sustainable development' is usually considered non-operational. This, however, underestimates the potential of the concept as proposed by the WCED. Based on the two ethically based normative ideas of not overburdening the carrying capacity of the Earth and a human right to equitable resource use, the claim is that we are producing too little wealth from too many resources (constituting an efficiency gap, and a justice gap for future generations), and that we are redistributing the wealth created too unevenly (a justice gap in this generation). Already from the need to overcome these two justice deficits, a number of perspectives and even quantitative policy goals can be derived, not by means of scientific proof, but by scientifically informed reasoning. This chapter tries to work out some fundamental implications of these ideas for the economy, society at large and the future of labour.

Sustainability and justice

Essentially, the WCED's understanding of sustainability is based on the two normative assumptions of intra- and intergenerational distributional justice, that is, the need for

- intragenerational justice: equitable access to the world's resources as a human right to the use of resources and the common heritage of mankind, and

- intergenerational justice: the availability of equivalent services from the environment for future generations.

Considering these principles, the current situation is unsustainable in a twofold way. On the one hand, the distribution of wealth in and between countries exhibits growing disparities. On the other, we are at, and in some respects already beyond, the limits of the Earth's carrying capacity, and only the inertia of complex systems has prevented more visible distortions of our living conditions than increased UV-β radiation, storms, floods, depleted fish stocks and so on already provide. Obviously, besides establishing a more equitable pattern of distribution, we have to limit (and indeed to reduce, given the damage already visible) the overall environmental distortions caused by our economies.

In order to reduce the overall impact of economic activity on the environment, we cannot focus on specific symptoms or well-known cause/effect relations, but have to reduce the total entropy generation stemming from the resource depletion caused by our economies. Due to prevailing ignorance, the necessary reduction of resources should be based on an extended precautionary principle. This will limit the applicability of cost–benefit analysis, so new ways of assessing policy measures need to be found to achieve sustainability. (See Figure 2.1)

The work of the Wuppertal Institute in Germany and the Rocky Mountain Institute in the USA has shown how it is possible to increase 'resource productivity', that is the amount of value (e.g. warmth, cooling, lighting, mobility, comfort) we derive from a given amount of resources. Thus the book *Factor Four* (Weizsäcker, 1995) is subtitled 'Doubling wealth, halving resource use'. Factor four here means getting

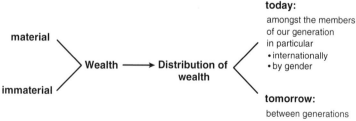

Figure 2.1   Sustainability is a matter of justice

the same amount of value for one quarter the resource inputs. As a first target, reduction of global resource consumption by half has been proposed for energy and for material flows (Schmidt-Bleek, 1992: 40–5; 1993: 456–62; for further sources see Spangenberg et al., 1999: 10–12). For land use, only qualitative estimates exist. Taking into account the justice principles outlined above, it is an ethical imperative for humankind to share the use of limited resources in a more equitable way.

In the European average, fair distribution suggests a need to reduce energy consumption to one quarter present levels (a factor of 4), material input to one tenth (a factor of 10) and land use to about two-thirds present levels. For the South, fair distribution within the permissible consumption limits suggests, again on average, a *doubling* of resource availability compared to current standards. This is what is called 'living in our environmental space'. The environmental space available for the inhabitants of Europe as a whole (Spangenberg, 1995: 16–117) and for 31 individual countries has been calculated in the course of the project 'Towards Sustainable Europe' (Carley and Spapens, 1998). Obviously, living within our environmental space is not only a problem of physical resources available, but cannot be achieved without closing or at least narrowing the income gaps between rich and poor nations and groups.

Sustainable economics

No need for a decrease in material wellbeing is detectable from the reduction targets for materials, energy and land given in our calculations, since these targets can be reached whilst maintaining a constant amount of services (Weizsäcker et al., 1996). Nonetheless, impacts on the economy and on lifestyles will be significant. In order to assess the impacts on economies and societies, the dematerialization target of reducing resource extraction from the environment to one-tenth of current levels has to be analysed. This analysis must take note of the impact of such a reduction on technology demand, consumption and production patterns, competitiveness, and in particular on growth, since growth is the traditional, supposedly painfree answer to all distributional questions, as even the Brundtland report illustrates.

Since the necessary dematerialization by a factor of ten in the next 50 years was measured in absolute terms, the factor will increase in the case of a growing economy in order to keep the throughput of materials at the environmentally justifiable level. Growth in this context is understood in the usual sense of GDP growth, since this is suitable to measure the financial turnover of national economies and thus to characterize the dematerialization needs for a stabilized resource throughput in terms of tonnes per euro or US$. However, this is not intended to imply that GNP/GDP could be read as characterizing real wealth (Ekins and Max-Neef, 1992) or – even more misguided – wellbeing (van Dieren, 1995).

So, with an annual growth rate of 2 per cent, the necessary factor of dematerialization will be 27 (reduction of resource use to one twenty-seventh present levels), and with an annual growth rate of 3 per cent it will be a factor of 45. Whereas a reduction of throughput by a factor of ten may be technically feasible within 50 years, a reduction by a factor of 45 (or even by a factor of 200 within the next century) probably will not be.

Thus, the limits to material flows based on ethical concerns translate into limits to economic growth: even after reaching dematerialization by a factor of ten, growth must be limited to a maximum equal to the annual increase in resource productivity. This, however, means that, although annual economic growth for a couple of decades will be boosted by the necessary restructuring of the Western economies, in the long run the level of the annual increase of resource productivity forms an absolute ceiling on growth. For Central and Eastern Europe, where a fundamental restructuring of the economy is already under way, there is a need to give the transformation a new, sustainable direction and thus to create economic structures that are competitive. Restructuring for sustainability also demands the creation of jobs which are available not only in the short run, but in a long-term perspective. For the South, plagued with the IMF/World Bank structural adjustment programmes it means insisting on giving the structural adjustment a sustainable direction instead of primarily orienting it towards globalization, export earnings and debt service. Some of the rethinking triggered by the Asian financial crisis may provide solid starting points for such a process.

The current patterns of growth, however, are environmentally disruptive, macroeconomically counterproductive, and socially divisive, with widespread unemployment and polarization between rich and poor. Obviously, growth as such cannot solve our problems.

Sustainable societies

Sooner or later we will have to live within limits, an insight that for many people will require significant changes in their value systems. In that case, although individual companies may still grow at the expense of their respective competitors, for the business sector as a whole the possible gains from economic growth will be limited. It is plausible to assume that at least the 'losers' among companies will claim the benefits from productivity increase for themselves, instead of sharing them with their staff. This would decrease the amount of finance to be distributed among the employees. The perceived contradiction of competitiveness and sustainability is based on the prevailing narrow and short-term understanding of competitiveness. This understanding needs to be replaced by a broader approach including the meso- and metalevel elements of competitiveness as well as consideration of possible gains from social security and quality of life (Messner, 1995; Hinterberger et al., 1997).

Consequently, the implementation of policies towards sustainability will probably be hardest for those societies already facing the most severe inequities. These countries, where there is no social climate of burden-sharing, include (besides a number of developing and transition countries and Switzerland) the Anglo-Saxon nations. It may be no coincidence that those countries with the lowest inequality ratio (which often comes with a more consensual political tradition) frequently have a high profile in national and international environmental affairs, like the Scandinavian countries or the Netherlands.

The absolute and relative income of the richest 20 per cent of the population has been increasing considerably over the last 15 years in most OECD countries, contributing not only to growing income differentials but also to very expensive life styles and wasteful consumption patterns. For a move towards sustainability, these tendencies need to be reversed. Reduction of resource use must go together with more equitable distribution patterns – or it will contribute to increasing poverty, and so will fail, in ethical as well as in political terms.

The transformation towards sustainability will necessarily – like all fundamental transformations – cause severe social tensions, and if it is to find public acceptance, it must contain a strong component of increased distributional justice. Thus besides being an integral part of the very definition of sustainability, decreasing instead of increasing income disparities is a crucial precondition for a broad acceptance of any sustainability strategy and thus for its success in Europe. The rapidly developing poverty in large parts of the South and in Eastern Europe, and the confrontation with the small class of thriving capitalists in these countries poses a specific risk for any transformation towards sustainability. In the South these considerations mean that combating poverty must be given a high priority in all national development strategies, instead of focusing on capital intensive industrial development.

The potential for efficient technological innovations in private transport suggests that resource efficiency targets can be met. Goods transport could be doubled in efficiency, which with reduced transport volumes due to dematerialization and changes in the modal split makes a factor 10 reduction possible as well. Politically, these developments need new institutions, in particular a combination of 'pull' measures (attractive alternatives) and 'push' ones (in particular increased transport cost). With reduced throughput and increasing transport cost per tonne and kilometre, international trade will also be gradually reduced and restructured. Whereas today the majority of material transports are bulk materials (mainly raw materials), increasing transport expenditure will justify long-range transport only for those goods that have a significant added value. This does not necessarily mean a decreasing value of trade or a reduced income from it, but the restructuring of global trade towards the exchange of processed goods instead of raw materials, as is already the case among OECD countries at large. Such a restructuring of markets, however, implies a stronger role for local and regional economic structures by extending the range of regional products and services.

The future of labour

In a sustainable society, paid labour must in its basic organization reflect the principles of sustainability. This is a task for management and company owners, and will need self-organization processes and contributions of trade unions as well. These are the key players in any operable sustainability strategy. From an ethical point of view, the need for the fair sharing of labour between employed and unemployed, for example, by means of reduced working hours is of central importance.

A Western society aiming for sustainability can probably achieve higher employment, if it chooses socio-ecological tax reform as a key tool. In this case, additional revenue is generated by 'getting the prices right'. Job creation is created by redistribution. The most promising way of doing this is a combination of reducing the labour cost of low productivity work, as varied as environmental protection, the arts, or nursing (it says something about the value system of our societies that 'caring' for cars is considered more productive and is better paid than caring for children or the elderly). This needs to be combined with investment programmes to update the infrastructures of our societies, which at least in EU countries can be financed from redirecting and phasing out of public subsidies.

Although environmental politics is no substitute for labour politics and its successes must not be measured in terms of jobs created, there will be a significant impact, at least in Western Europe. Since this will decrease the pressure on public budgets due to decreasing social spending, the resulting surplus should be used to stabilize pension and social security systems and to decrease the public debt. For pensions, a solution based on the existing publicly guaranteed insurance systems would be most appropriate, since a shift to financing pensions from shares and bonds would, for example in Germany, require a tripling of the total capital stock of the national economy by 2030 in order to provide a sufficient financial volume at current revenue rates. This is a condition that would hardly be compatible with a dematerialization strategy. Reduction of the public debt is an issue of intergenerational justice, and it will be necessary to make sure that political intervention is still possible once public income from taxation stops growing as GDP growth slows down due to the reduced raw material throughput. The latter, however, will take thirty years or more, leaving appropriate time for all necessary adjustments. (See Figure 2.2)

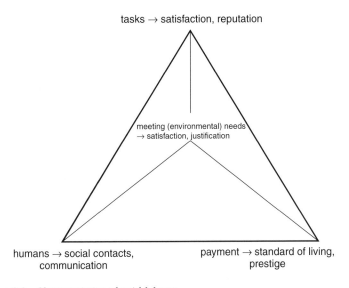

tasks → satisfaction, reputation

meeting (environmental) needs → satisfaction, justification

humans → social contacts, communication

payment → standard of living, prestige

Figure 2.2 Characteristics of paid labour

While increase in transport cost would protect production, services and jobs from international competition, it would also constitute a division of national economies between one sector with high competition, high productivity and high salaries, and another sector with probably more safe jobs but fewer well paid ones. Balancing income levels and making the move from one sector to the other possible will be one of the key tasks of future social and labour politics in order to realize the ethical principle of equity of opportunity. In the 'South' (countries with low per capita consumption), however, a sustainability policy would not automatically increase employment levels. On the one hand, the decreasing demand for raw materials from the 'North' (high per capita consumption countries) would create hardships for the export business (usually not the poorest group in a country). On the other hand, however, it would provide a better chance to use resources for development of the national economy and for a better standard of living for the poor. The weakening grip of global competition on local markets may, furthermore, open windows of opportunity for a more labour-intensive industrialization strategy. However, for this to happen, a new social contract would be necessary to combat poverty by empowerment and more distributional justice, including land reform, access to resources and so on. This would be an important element of any national development planning. If national policy elites do not exhibit leadership towards these ends, the opportunity may well be lost.

Not only do the number and distribution of jobs need to change, but the organization of labour as well. Whereas today the highest productivity and maximum efficiency is usually found in industrially organized labour, satisfaction from labour is more evident in self-organized work in household and community (Scherhorn, 1993: 17–23). So one of the key tasks for the future is to bring more elements from private work into the factory, such as a balance between autonomy and teamwork, and fewer hierarchies. And the subsistence or self-organized sector should take advantage of elements from formal labour, for example, safety standards, accident and health insurance, and access to social security schemes.

These changes will promote job satisfaction and innovation in the labour force. Although this may sound like socialism, it has led, for example in the German automobile industry, to the highest salaries and shortest working hours in international comparison, and to the highest productivity and competitiveness. Consequently, in continental Europe similar measures have already been proposed by business consultants with no view to sustainability, merely to increase the competitiveness of companies, as one step beyond lean management concepts. This change of the basic organizational characteristics of industrial labour is also a key precondition for the development of sustainable consumption patterns. Today in Europe, significant and dynamically growing purchases are those motivated by 'compensatory consumption' of status symbol goods: people spending money they do not have on things they do not need to impress people they do not like. The preconditions for more sustainable consumption patterns linked to labour are exactly those described previously, for reasons of competitiveness and sustainable production.

Sustainable lifestyles

The prevailing global consumption pattern is based on the European lifestyle, which was spread during the colonial period, enforced by the world economy, and driven to extremes by the economic elites in the US and some Third World countries. Today, those social strata are forming the Global Middle Class, which is setting the standards for the pursuit of happiness, defining what is regarded as a satisfying life and thus driving the development of consumption aspirations all over the world. An analysis of per capita energy consumption in San Diego, USA, showed that the wealthiest households spent three times more money on energy than the poorest ones, but used 5.33 times more energy than poor households. The difference, a factor of five, is approximately the same as between the average citizens of the USA and Argentina (*Ökologische Briefe*, 1995: 7) and indicates a lack of social justice as well as of environmental sustainability in affluent societies.

In contrast to these lifestyles, sustainable consumption is based on a simple idea: it is not the quantity of ownership that counts for the quality of life, but the quality and quantity of accessible services. This breaks the conceptual links between quantity and quality as well as between owning and using, and consequently permits a new definition of sustainable wealth: availability of a high level of quality services, while reducing the throughput of the economy. A change in consumer demand towards this view, partly based on ethical concerns and supported by eco-tax reform, would also play a crucial role in restructuring business. So far, no company has ever managed to survive against the will of its customers in the long run. Therefore business will have to adapt to changing priorities, including the preference for values like 'better' rather than 'more', and for 'being' instead of 'having' once household consumption directs them this way. However, the influence of households is more limited than economic theory suggests, and is significant only in the fields of construction and housing, nutrition and mobility (Lorek and Spangenberg, 1999).

With changing resource intensity the products will necessarily change as well. Products for sustainability will be less resource intensive, more durable and repairable, and will need less work for production, but more for maintenance, repair and recycling. The significance of these processes depends on the range of products involved, which in turn is influenced by societal institutions such as housing patterns, family size, demand for living space per capita and a generally accepted definition of what are desirable status goods. These products might be more expensive, to justify the salaries for repeated repair and maintenance, but their extended life spans would decrease the price per service gained from them. (See Figure 2.3)

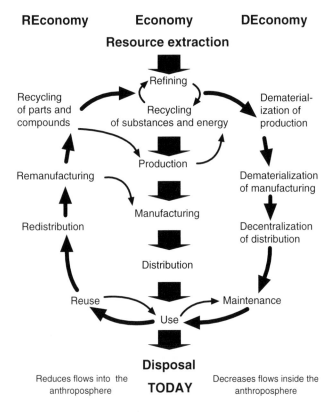

Figure 2.3   From Economy to REconomy and DEconomy – material flows in a sustainable society

One strategy to overcome supply limitations due to increased relative product prices (as compared to salaries) is the sharing of goods that are not in permanent individual use, thus providing access to more services while releasing a burden from individual budgets as well as from the environment. Housing patterns and family structures that support this kind of consumption would at the same time extend the market for such products, in particular for housing and cars.

As well as these changes in the quality and quantity of production and consumption, the spatial structure of our living and the organizational pattern of sustainable societies will have to be quite different from today's. In all agricultural and forest areas, where – besides the extension of protected areas to about 10 per cent of the landscape – a transition to organic agriculture and sustainable forestry is needed, this will cause severe problems including the balancing of individual ownership rights against public demands. On the other hand, the phasing out of transport growth and strengthened supply from regional sources (for cost reasons, as mentioned earlier) will add to declining demand for additional land for transport. Some sectors, such as food importing or long-distance tourism, would suffer from this development, whereas others, including domestic tourism, regional food and beverage production, would gain.

As a prerequisite, education and qualification patterns will not only have to be different from today's, but will need to be highly dynamic – life-long learning will become more important than ever, and ethics will have to play a more dominant role.

Sustainability politics

Sustainability, as defined here, has provoked an intense dispute as well as highly varying judgements about its relationship to 'traditional' political groupings and categories. Hardly any other concept has been accompanied by such varying judgements: labelled a new kind of central planning approach by some business associations, some major national and transnational companies have welcomed sustainability as an inspiring new way of thought, creating immense business opportunities. Whereas some traditional left groupings have seen a mere greenwashing of capitalism, others have found it to be a new paradigm for the left. So, at the end of the day, what is it really?

First, it is a new, integrated, ethically based normative system of rules (in sociology called institutions). Its basic values, however, are not new at all. Concepts like the service economy or the consumer society, value-based policy goals like democratizing labour, participation within civil society, conserving the common heritage of humankind and quality of life, and, in particular, ethical norms like freedom, equity and solidarity have been guiding European policy debates over the past two centuries, since the French Revolution. Sustainability just provides a unifying framework for many old and new ideas. This, however, is not a weakness but a strength – it can draw upon the experiences of past struggles and lessons learnt. In this sense, it can become a paradigm, not by recruiting followers of its own, but by providing a framework for people active for a better society in different places on different issues.

Although there can be no denying that there is significant room for improvement of the dominating global capitalist model, particularly in social and environmental concerns, there is fierce dispute about what kind of measures need to be taken. The ethical priority behind the sustainability paradigm then puts justice and the satisfaction of human needs ahead of market values like business efficiency and profitability. On this basis, private business, public authorities, non-government organizations, trade unions and others can contribute to the shared perspective according to their own capabilities. In this sense it is consensus-oriented, based on stakeholder integration and shared responsibility.

Secondly, since the concept has been worked out to change the status quo, and since the status quo is based on global capitalism, it is unavoidably critical of the current capitalism, although some people label it 'enlightened capitalism' (Carley and Spapens, 1998). Its critique, however, does not copy any traditional approach: criteria and alternatives are very different. (See Figure 2.4)

Thirdly, the use of elements of leftist thinking, selectively based on past experience, is accompanied by a similarly selective approach towards liberal and conservative ideas, but based on a common set of norms and ethical criteria. For example, the promotion of democracy and human rights is from its origins a liberal one, as is the emphasis on the market economy. The market, however, is envisaged to be complemented by institutions to enhance distributional justice (Daly, 1991). In Europe, this approach is a respected policy, developed after the Second World War and mainly based on the 'ordoliberal' economic theory (see Renner, 1998) of the Catholic and the reformed church, and the so-called social market economy (see Kirchenamt der Evangelischen Kirche in Deutschland, 1997).

Sustainability has conservative roots as well: the conservation of natural and cultural heritage reflects the momentum of conservative and

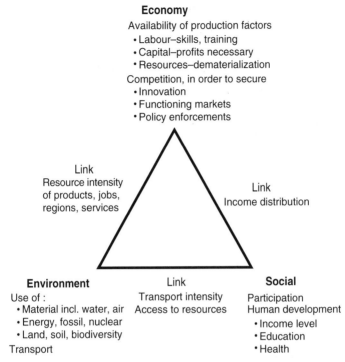

Figure 2.4   Taking account of the interlinkages is crucial for sustainability policies

even romantic thinking, based on respect for humans and nature. These traditions of thought are complemented by progressive elements like target setting, land use planning and so on, commonplace all over Europe, and attributed at least to some degree to progressive social-democratic thinking.

Altogether, sustainability is no new ideology, has no blueprints for future societies, but has some ethically based criteria for the quality of future life, based on the need for solidarity within and between generations. Following these criteria would enhance the quality of life for the majority of people, and offer a dignified life to all citizens. It would promote democracy, transparency and participation, thus combating bribery and personal dependencies. Business people, administrators, trade unionists, human rights activists and environmentalists together can really get more things going than has been possible before. This, however, needs cooperation and an attitude of non-confrontational problem-solving, and thus changes in the rules of the political game, the institutions of society. Some of these changes are already beginning to materialize, others have to be supported and promoted by information and awareness raising. In the long run, this could lead to an 'enlightened selfishness' (Sachs et al., 1998). Other changes like the introduction of proper incentives require political leadership.

New instruments are needed, but how can we handle their complexity?

Politics could be said to be decision-making based on incomplete information. The more incomplete the information, the higher obviously is the risk of failure. Consequently, managing the multifold complexity of the environment–society–economy interaction by central decision-making is bound to fail economically, socially and environmentally, since not all the relevant information is available, nor is the relevance of the existing information obvious. Nature is no simple mechanical system that is predictable and manageable.

Given this insight, what can politics do, if direct intervention and steering have a high risk of counterproductive (and all too often counterintuitive) effects? We propose a new way of thinking: more political responsibility with less intervention, setting framework conditions and letting the self-organizing dynamics work, while being led by the framework into a desired direction. Instruments like taxation, subsidies, land reform, grants and permissions, systems to improve income distribution, equal access to legal advice, and so on could give the economic dynamics a direction without interfering too much with day-to-day decision-making.

Directions must be based on broad consultation set by the legitimate government. Tools to effect change include new economic models, less one-sided than the currently prevailing neoclassical ones, including the role of labour, the value of nature, demand as well as supply-side effects and the changing needs and preferences of people. Indicators will help assess policy measures once a direction has been determined. These indicators can increase transparency and thus accountability, but are no substitute for detailed policy development.

Conclusion: Time enough, but no time to lose

Dramatic changes are foreseeable in the scenarios presented, but they are not more dramatic than those we have seen in the past fifty years: the post-1945 rebuilding of Europe, the changes in Japan and China, the communist block and the colonial empires. Fifty years ago the term 'development' was not born, nor 'sustainability', nor 'environment'.

Today, closing the poverty gap in and between societies, reducing resource consumption, and modernizing the economy are goals which sound ambitious, but the changes in the next fifty years can hardly be greater than those we have been through in the last. Sustainability is not a mission impossible, but a vision impossible to ignore.

3
The Intergovernmental Panel on Climate Change: Beyond Monitoring?

Elizabeth Edmondson

Introduction

The creation of effective climate change regimes has become essential, and increasingly urgent, in the face of continuing difficulties in reducing the global production and consumption of greenhouse and ozone-depleting gases in the 1990s. The international community, predominantly comprised of states' delegates and scientists, has agreed several times over that greater knowledge and understanding of climate change processes and consequences will assist the development and implementation of policies to minimize the social, economic and environmental impact of climate change (Greene, 1996: 196). As part of these negotiations, parties have revisited some of the difficulties of allocating responsibility for global environmental protection, international security, and the elimination of poverty and initiatives to promote international governance. Climate change management strategies raise ethical questions concerning relative industrial capacities and the distribution of consequences, responsibilities and burdens within the international political system.

The history of the Intergovernmental Panel on Climate Change (IPCC) activities show it to be a flexible institution, likely to proceed incrementally in order to maximize agreements between parties. This is crucial since it increases the likelihood of compliance. It also increases the likelihood that agreements will be fair (or be deemed fair enough by a majority of parties, including those most directly affected), take fuller account of multiple policy interests within states, and promote a multidimensional policy approach. Lasting negotiating coalitions within the IPCC and established scientific and policy-making networks suggest that, to date, agreements have been fair (if slow). Indeed, even strong critics of the IPCC acknowledge its commitment to participatory decision-making, informed policy-making and efforts to accommodate diverse and sometimes competing interests among actors (Mintzer and Leonard, 1994: 322).

If the IPCC is to achieve a role in compliance and verification – and long-term climate change management, including effective emission reductions, almost certainly depends upon this – then, existing structural features of the IPCC itself, as well as the international political system, must be used to advantage by policy-makers. Those best positioned to pursue policies and collective action strategies are those whose roles within the wide array of environmental regimes and agencies and other international institutions are well established. Consequently, deliberate and reflective attention to institutional linkages between the IPCC, the United Nations Environment Program (UNEP) and the World Meteorological Organization (WMO) seems likely to provide fertile ground for policy development and effective international climate change regimes. This approach is reliant upon the commitments of states and continued contribution of knowledge-based networks. This derives from the impact of knowledge, especially the accumulation of scientific knowledge and the generation of credible predictions of climate change consequences as a basis for international policy-making.

Climate change regimes are more reliant upon international law (and corresponding domestic legal frameworks) than upon other environmental problems which have achieved management strategies, such as fisheries, deep seabed mining, Antarctica and transboundary pollution (Sebenius, 1994: 315). International climate change regimes have taken longer to be created and have involved unprecedented levels of negotiation and consultation between states and other actors. In climate change management, ethical issues cannot be sidestepped or given diminished status. Highly industrialized states are now widely identified as those most responsible for impending climate change. The most dire consequences of unmanaged climate change – loss of territory, disruptions to climatic patterns, desertification and so on – are expected to have greater effects on less-developed states. Unlikely as it might seem in a political system characterized by the competing interests of states, the international community is beginning to agree that managed climate change is the responsibility of all states, since sustainable development is only possible when pursued globally.

This chapter argues that the IPCC acts as a central pillar in the creation of climate change regimes, taking account of a corresponding urgent need for burden sharing towards sustainable development. The level of knowledge and communication networks within the IPCC, together with its existence as an integrated, international institution, set it neatly at the centre of negotiations towards comprehensive regimes. Throughout this chapter it is argued that the structures and functions of the IPCC equip it for an expanded role as manager of international agreements. To this end, an account is given of its operating structures, management practices and links with other climate change and sustainable development bodies. The chapter also considers the characteristics of the IPCC and seeks to identify ways in which these promote effective agreements: that is, facilitate comprehensive regimes.

The IPCC – structure, practices, context

Membership and organizational structure

One of the most remarkable things about the IPCC is its extensive and diverse membership, which exceeds 190 states, transcends regional and state boundaries, and has attracted broad support from within the UN General Assembly. Although the majority of scientific experts engaged in its research and monitoring activities are academics or research scientists working within state-supported research bodies, it is by no means dominated by industrial states. IPCC scientific activity derives from a diverse array of states: scientists from more than 50 countries play leading roles in IPCC analysis and communication.

Diverse states have been active in coordinating or managing Working Group research efforts: China, Argentina and Sierra Leone currently co-chair Working Groups in conjunction with the United Kingdom, the USA and the Netherlands. Scientific leadership and peer review is provided by an array of non-Western and developing states. For instance, India, Nepal, Mexico, Jamaica, Thailand, Tanzania, Bangladesh, Mauritius and Benin have all made leading scientific contributions. While members of the IPCC experience marked disparities in their domestic circumstances, as well as in their levels of power and influence within the international political system, a strong level of overlap exists in their interests concerning climate change.

The IPCC, which operates under the auspices of the WMO and the UNEP, was formed in 1988 by the international community acting through the United Nations General Assembly (Resolution 43/53).

This represented an effort to create a central institution through which diverse climate monitoring bodies and scientific research work throughout the world might be coordinated and their findings made available internationally to provide states with information acquired beyond their borders and, in many instances, beyond their own research capabilities. It is unlikely that the IPCC would have been formed without the initiatives of experts and scientists whose knowledge of the processes and impacts of climate change led them to believe that only international strategies to diminish impacts or reduce likely rates of change held any prospect of achieving these objectives (Boehmer-Christiansen, 1996: 175; Paterson, 1996: 143–4).

The functional separation of the Working Groups from the central Bureau and the Plenary Sessions of the IPCC (Figure 3.1) enables a partial separation of scientific research from policy-making which is crucial in establishing research parameters, identifying issues for policy negotiation and monitoring impacts. This is important to IPCC recommendations and strategies because it enables scientists and policymakers to assume relatively equal positions in negotiating processes. The IPCC Bureau comprises Working Group Co-chairs and Vice-chairs, each of them experts in aspects of climate change factors, processes or consequences. In addition, six Regional Representatives are also given places within the central Bureau, each nominated by IPCC government representatives of the relevant region.

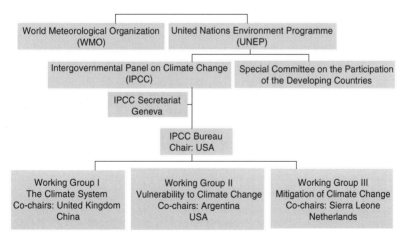

Figure 3.1 Organizational structure of the IPCC

The inclusion of Regional Representatives provides clear evidence of support within the IPCC for developing comprehensive climate change regimes that are cognizant of diversity in states' needs and consonant with their capacities to address climate change issues. It also highlights the parallel strands of scientific and state actors within the IPCC. These features reflect awareness within the IPCC of the importance of equity in sharing the costs of climate change management and reinforce commitment to equal status between diverse actors and sectors within negotiations. In attempting to establish comprehensive climate change regimes, parties seek to prevent the most severe climate change predictions without dooming the developing states to permanent poverty.

Working Groups – history and practices

The IPCC has adopted a structure of three Working Groups since its establishment, although the activities and foci of Working Groups have twice been re-configured (Figure 3.2). In each case, reorganization has coincided with frustration at slow progress towards comprehensive, effective, climate change regimes (Keohane, 1996: 19). Revisions to Working Group activities have also coincided with the identification of gaps in knowledge or data collection. Each alteration has afforded renewed hope of finding an appropriate organizational structure to maximize participation in monitoring programmes and efforts to generate lasting international agreements as solutions. Such efforts might be criticized as putting too much effort in the wrong place by paying more attention to finding the right organizational structure when it would be better to achieve agreed targets and ways of ensuring that

	Date	Task areas
Working Group I	1988	Scientific assessment
	1992	Scientific assessment
	1997	The climate system
Working Group II	1988	Impact assessment
	1992	Sectoral study groups
	1997	Vulnerability to climate change
Working Group III	1988	Response strategies
	1992	Socio-economic impacts of response strategies
	1997	Scientific, technical, environmental, economic and social aspects of mitigation

Figure 3.2   Working Group research areas

these are met. There is, however, much to be encouraged by in this approach by participants in the IPCC.

Finding the right organizational structure may well assist in the creation of effective regimes since negotiating coalitions established during structural revision are likely to coincide with the identification of shared interests, visions and implementing capacities. Institutional linkages between international climate change bodies and institutions for economic development reform enable the implementation of strategies to minimize the impact of climate change and to reduce greenhouse gas emissions. It is imperative that negotiating coalitions in climate change institutions form across policy sectors because this maximizes both the likelihood of achieving relatively fair distributions of direct and indirect costs and burdens, and establishing comprehensive regimes.

The most recent Working Group reconfiguration reveals a shift away from the sectoral approach to research and analysis adopted between 1992 and 1997. This signifies less that the sectoral approach failed or was of limited utility and more that participants are now more keenly committed to achieving progress towards analyses of the socio-economic consequences of climate change and possible minimization strategies. This is particularly significant since one of the greatest obstacles to the formation of comprehensive climate change regimes has concerned the relationship between climate change, greenhouse gases, industrial capacity, economic growth and political power or influence within the international system. This foregrounds the North–South debate concerning the distribution of burdens, blame and responsibilities for greenhouse emissions, and reform of the international political economy.

The operations of the IPCC reveal an awareness not only of the need for more and better knowledge concerning climate change factors and implications, but also of divergent approaches to the accumulation of such knowledge. This ensures that information is widely acquired and disseminated, and reinforces processes of expert peer (and cross-sectoral, diverse actor) review within the IPCC. The reasons, purposes and methods of acquiring knowledge vary widely between different actors, including scientific experts and policy-makers. Each of the research teams contributing to the findings of Working Groups has been permitted to select and adopt their own methodologies, targeted research area and modelling/projection devices. This is not to say that the Working Group structure encourages an *ad hoc* or loosely integrated approach to scientific research. Rather, it reveals something of the flexibility of the IPCC in accommodating and incorporating diverse perspectives, needs and contributions from its members.

Context of IPCC activity

The IPCC operates within the framework provided by both formal institutional links with other global research networks and monitoring programmes and informal links provided by Working Group scientists, diplomats and state delegates. Three global research networks have played prominent roles in IPCC activities: namely, the World Climate Research Programme (WCRP), the International Geosphere Biosphere Programme (IGBP), and the Human Dimensions Programme (HDP). The IPCC has been supported in its monitoring operations by links with the Global Climate Observing System (GCOS), the Global Terrestrial Observing System (GTOS), the Global Ocean Observing System (GOOS), World Weather Watch (WWW) and Global Atmospheric Watch (GAW).[1] World Weather Watch provides vital links, structural support and an ongoing source of information to the IPCC at the levels of Working Group I and the IPCC Bureau. Most notably, it provides established international data collection through its network of 9500 land-based observation stations (the Global Observing System), and a centralized data collection, compilation and analytical capacity (Soroos and Nikitina, 1995: 73).

It was clear from the international negotiations that had resulted in early greenhouse gas and climate change agreements that the majority of states were keen to engage in assessments of climate change consequences and possible solutions/alleviation strategies. In this regard, the IPCC provides an organizational mechanism through which long-term cooperation and communication produces established coalitions of interest, consolidates specialized expertise which is uniquely cross-sectoral and promotes policies – programmes of collective action – with clearly defined targets and objectives. Broadly based international negotiation results in non-discriminatory policy advice to states, encourages technological transfers and renews enthusiasm for achieving targets. The setting of new goals at the 1998 Buenos Aires Framework Convention on Climate Change (FCCC) Conference of Parties – the elaboration of the Plan of Action – and the decision-making of the fourteenth session of the Plenary of the IPCC (Vienna, 1–3 October 1998) in identifying Lead Authors, Contributors and Reviewers for the Third Assessment Report (TAR) indicate a high degree of commitment to achieving effective agreements and mechanisms to ensure their ready and general implementation.

Increased support for the Global Environment Facility (GEF) and the inclusion of technology transfers and emissions trading reveal heightened efforts in international climate change institutions to address the interests of developing countries. While technology transfers and emissions trading provide relatively weak levels of protection for the environment, they offer mechanisms by which the economic burdens of responsible development might be shared between rich and poor states. Importantly, these mechanisms have also reintroduced notions of a global commons into continuing efforts to create comprehensive climate change regimes. They are supportive of efforts towards multilateral burden sharing and the allocation of cross-sectoral, multi-dimensional, responsibilities in climate change management.

Effective agreements

Negotiating effective agreements

Effective institutions are achieved only by close international agreement, and the maintenance of agreements remains permanently dependent upon parties having confidence in the management structures of organizations and institutions. Effective climate change regimes are heavily reliant upon the organizational structures of the IPCC and the FCCC, their shared knowledge, parties and associated knowledge-based communities (Young, 1994; Soroos, 1997: 22). In the case of the IPCC, the Working Groups are the most crucial elements since these provide the data, projections and recommendations from which all other climate change management bodies derive information, both within and beyond the IPCC. In the case of the FCCC, the structures and focus of the Secretariat are most crucial since these are linked in setting priorities for the GEF, through which fuller attention might be given to issues of sustainability and cross-sectoral impacts. Combined, these institutions provide a likely mechanism through which ozone depleting and greenhouse gas emissions targets might be matched with industrial capacity, growth targets, and, possibly, distribution of social costs.

A degree of compatibility and mutual interest is essential for effective implementation of collective action strategies, especially where conflicting or complex networks of interests and actors are present (Baum and Oliver, 1991; Connolly, 1996: 334). Such strategies are more likely to be implemented rapidly and effectively when the objectives sought coincide with a broader range of objectives held by participants, and where participants are confident that the institutions they have created will meet their objectives (Young, 1994: 107–14). In the case of the

IPCC, the Working Group structure enables participants to isolate aspects of climate change processes and impacts, thereby providing simplified, accessible information about key areas. This affords considerable opportunity for scientific networks to exert influence over policy decisions, problem identification and definition, and enables them to act as gatekeepers to knowledge accumulation and information dissemination. While this is intrinsically problematic for those who seek politically neutral knowledge, it is not especially problematic in the realm of international political affairs where actors are accustomed to bargaining, establishing agenda and manipulating interests.

Promoting international environmental agreements through negotiations within institutions such as the IPCC is inevitably an exercise in political persuasion and influence. Trade-offs, compromise, bargaining, promoting certain interests at the expense of other interests are inherent to the negotiating processes and forums within which they occur and the agreements or outcomes they produce. Knowledge, interests, priorities, identification of need, bargaining, creating managing bodies and so on are all thoroughly political enterprises. An institution that places scientific networks and their communication of knowledge (including its predictive capacities) at its centre, in the manner of the IPCC, remains a political institution. It is this feature – the marriage between scientific knowledge and policy-making via cross-sectoral activities and communication – that gives the IPCC a crucial role in developing negotiated climate change regimes.

The institutional character of the IPCC is essential in ongoing research, analysis and communication towards negotiated international climate change regimes. Among its features are a strong trust-building capacity that derives from shared activities and expectations to reinforce reciprocal, multilateral cooperation. It is important, though, to bear in mind that such trust is built upon common knowledge of climate change consequences and a mutual need to diminish climate change impacts (Paterson, 1996: 141). Consensual decision-making, cross-sectoral research, data compilation and modelling within the IPCC work towards strong levels of participation and communication between actors.

Role of knowledge in policy-making

The consensual decision-making approach offered by epistemic communities[2] within the IPCC may be regarded as an appropriate reflective response to the imperatives of multiactor, multidimensional objectives of participating state delegates. The nature of decisions to be taken by state actors within international forums necessitates the building of mutual trust via adherence to agreed principles of fairness (as well as expediency) and regular opportunities to review and amend decisions. Consensual decision-making within such forums facilitates agreement and effective implementation, especially where states hold diverse interests and are provided with opportunities to take new decisions. Epistemic communities aid the formation and implementation of international environmental policy by encouraging consensual decision-making within international forums (Feldman, 1995).

Epistemic communities within the IPCC provide an essential sharing of knowledge which enables parties to establish a degree of comparability in developing possible solutions to identified and mutually recognized shared problems. These processes are characterized by incremental decision-making within established guidelines that seek to accommodate as broad a range of interests as possible. Shared knowledge and vision between scientists and policy-makers has contributed to a narrowing of the gap between sectoral interests, creating stable negotiating coalitions and flexible agreements. Typically, amendments to policies, decision-making processes or structures follow new levels of shared knowledge.

The nature of epistemic communities has enabled them to develop particular roles within the international political system. Their existence as non-governmental networks of experts in fields relating to policy and management strategy development ensures that the information they provide is considered credible and, if not politically neutral, at least not merely politically motivated, nor driven by imperatives beyond what the knowledge itself implies. While the linkage between policy agreements or international regimes formation and epistemic communities generally is unclear, in the case of the IPCC these links are more apparent (Paterson, 1996: 140, 146). Clear information, communication networks and the processes of peer review within the IPCC promote trust between actors and enhance negotiations by providing clear parameters for discussion and bargaining.

These mechanisms encourage a bridging between scientific and political interests and actors, reinforce the pursuit of international objectives and aid ongoing negotiation and agreement. In this, the Bureau reflects the fullest interests and objectives of the IPCC, its auspicing bodies and the other institutions with which it is structurally and practically linked. It also ensures that the processes of expert peer review and consensual decision-making adopted within Working Groups and Plenary Sessions are also mirrored in the IPCC executive.

Knowledge, within this model, is treated in a manner reminiscent of individual freedoms in liberal democratic thought: it becomes one of the foundations upon which any pursuit of truth, explanation or policy development might occur. Without knowledge, or more accurately, without enough of the right kind of knowledge, states (the policy-makers) are doomed to make poor decisions even when they negotiate, and potentially even when they achieve regulatory mechanisms.

Effective negotiated agreements

Negotiated agreement and cooperation are central to the operations of the IPCC. This is evident in the activities of the Working Groups, the Bureau and Consultative Meetings. Without cooperation, negotiated multilateral agreements are destined to fail miserably, either because they need to be kept so general as to remain untargeted, or to be so highly focused or limited as to be isolated from other policy imperatives and dimensions. Alternatively, they may be established in such a manner as to render implementation, or implementation review processes, impossible or impracticable. For cooperation to be effective in international environmental management it must be organized (Soroos, 1997: 177), since organization is at the heart of the international political system. The plethora of environmental agreements which have been negotiated during the last two decades signify the importance of organized, institutionalized approaches (Greene, 1996: 196).

Negotiated agreements make good policy as a direct result of the trade-offs that parties make in reconciling their divergent or conflicting interests during negotiating processes. That parties are required to establish priorities of interest and to bargain in pursuit of protecting or enhancing identified priorities by collective declaration of interests provides for public and peer scrutiny. Negotiation and bargaining encourage parties to reflect upon their positions – not only in light of new or additional information or political pressure – but, most importantly, in terms of reconciling competing imperatives within their own policy-making spheres. Negotiated agreements not only maximize the likelihood of achieving collective action strategies that will be implemented, but also narrow the gap between states' international and domestic interests and activities. In addition, negotiated multilateral (collective action) agreements seek to be inclusive, to encompass the interests and priorities of as many parties as possible, because their effectiveness depends upon broad implementation.

Negotiated agreements to reduce the risks of climate change and justly distribute the costs of minimization strategies are vital. Negotiated agreements maximize compliance and assist the creation of enforcement and monitoring mechanisms because they facilitate cooperation between actors. Established negotiating frameworks add a sense of permanence that strengthens the credibility (and may also promote strong levels of compliance) of collective action agreements. These features coincide with the organizational characteristics of regimes and thereby serve to promote their creation and implementation (Levy et al., 1993; Keohane, 1996).

Beyond monitoring – compliance and verification

Expanding climate change management

There were promising indications at both the Thirteenth (Maldives, 21–28 September 1997) and Fourteenth (Vienna, 1–3 October 1998) Sessions of the IPCC that compliance has become an important issue in achieving agreements as well as in their implementation. Similar indications were also apparent at the third and fourth FCCC Conference of the Parties held at Kyoto, 1–10 December 1997 and Buenos Aires, 2–13 November 1998, respectively. Building on the Kyoto Protocol's 5 per cent emissions reduction target, the Buenos Aires Plan of Action seeks to consolidate the three central elements of the 1997 Protocol. These elements are: an international emissions trading regime which enables industrialized countries to buy and sell emissions credits; a clean development mechanism; and a joint implementation programme to provide credits for financing emissions-avoiding projects in developing and economic transition countries. Among other things, each of these initiatives seeks to establish auditing and verification criteria and defined institutional roles.

With the exception of widely and readily achieved emission reductions, the bulk of international agreements in the early 1990s focused on data-collection and long-term planning. While such a research-oriented approach provided governments with ready and credible reasons for avoiding specific or short-term commitments to environmental protection, intensive research was also essential in establishing comparable and collectively assessed information by which relative positions or commitments and verification procedures could be established. The introduction of verification procedures can limit the participation of states in international institutions or result in highly generalized agreements (Young, 1994; Greene, 1996). However, the incremental adoption of targets as key components of climate change and greenhouse gas emissions agreements is an important factor in maximizing effectiveness and compliance (Jager and O'Riordan, 1996).

Incremental targets provide continuing opportunities for implementation review, problem redefinition, participation by new parties and actors and the setting of new goals. These features are crucial to continuing participation in international negotiations and the formation of regimes. They offer strong prospects of compliance and the accommodation of diverse needs in target-setting. They also promote implementation review whereby scientific and state actors continue to share influence over decision-making, and where diverse states continue to renew their commitment to preventing unmanaged climate change and responsibility for the implementation of collective agreements.

Utilizing institutional linkages and specific targets, these steps may well represent the foundation of a comprehensive climate change regime whereby the historical obstacles to collective action (distribution of burden, unequal industrial capacity, and so on) might provide vital mechanisms for emissions reductions and sustainable development. Incremental targets allow for burden-sharing and industrial adjustment to promote long-term resource security: ecologically sustainable development and comprehensive environmental protection regimes might follow. Resource and territorial security remain vital common interests for states. In an atmosphere of negotiated trust and mutual interdependency, these state interests contribute to regime formation.

Criticism and caveats

It is evident that the IPCC is an elite body, composed as it is of large numbers of scientific experts. It is not, however, an *elitist* institution, seeking to create an insider group or exclusive club with 'members only' benefits. By contrast, it is highly inclusive in its membership, its relations with other institutions and individual actors. Most notable is its use of consensual decision-making practices. There are no membership weightings or special privileges evident in the structures or activities of the IPCC. The inclusion of Regional Representatives as members of the Bureau signifies an emphasis on equal access to information, opportunities to initiate and coordinate research and analysis, as well as ensuring full participation and evenly shared responsibilities within the IPCC.

The epistemic communities within the IPCC play crucial roles in determining IPCC activities. They are by no means apolitical or policy neutral actors and, to some extent, criticisms that the IPCC is, or has been, driven by scientists have some validity (Boehmer-Christiansen, 1996: 175). Notwithstanding such criticism, the essential character of the IPCC is as an international political institution. In this it shares institutional and organizational links with other entities and comprises a relatively static and permanent membership. It was established by political actors to gather scientific knowledge to enable those political actors to make informed choices concerning current and future economic and industrial activities, international relations, renewed security assessments and governance mechanisms (Paterson, 1996: 60–5).

The way forward

The establishment in 1997 of an IPCC Data Distribution Centre (DDC) following the Second Assessment Report of Working Group II (1994/5) and the founding of the FCCC Secretariat in Bonn in 1995 provide tangible evidence of a shift towards target-setting and compliance measures. The DDC seems likely to expand the international network of climate change researchers and the pool of knowledge from which projections (and policies) are created. In addition to providing accessible information and impacts assessments, the DDC also provides socio-economic data, matching the cross-sectoral, multidisciplinary model of Working Group activities. The models and structures of the IPCC offer a template for cross-sectoral approaches that are also reflective, seeking to develop comprehensive, integrated regimes that are equitable. This means regimes that are at least fair enough to attract broad levels of support and withstand expert and state actor scrutiny across several policy dimensions.

Regimes can only function to the extent that they are provided with institutional structures that enable the implementation of agreed actions, objective and functional review, monitoring of outcomes, reconciliation of diverse interests and the accumulation of new knowledge. Verification or compliance assessment will, almost certainly, become increasingly important in environmental regimes as distributive issues and consequences of negotiated agreements become matters for negotiation rather than potential log-jams in bargaining. Climate change management requires international governance and the only possible means of achieving this lies in the establishment and maintenance of regimes – institutions that embody shared values and knowledge to reinforce their legal bases. It is this aspect of regimes that makes the expansion of IPCC functions to include some form of compliance verification or target enforcement likely within the next cycle of negotiation, research and reporting.

The IPCC has taken significant steps towards the creation of comprehensive climate change regimes. It has initiated scientific and political meetings, coordinated international research, monitoring, data collection and analysis. By facilitating meetings it has sustained international attention within its membership and consolidated links with other climate change and environmental management institutions and organizations. The IPCC has facilitated negotiated agreements, established negotiating coalitions between scientific and policy-making actors and continued to reinvigorate interest in climate change problems. In this manner, the IPCC has displayed a self-perpetuating vision of integrating and coordinating scientific and political interests to minimize risks and impacts of global climate change.

Conclusion

Global knowledge of climate change processes and consequences would not have been achieved in the last decade in the absence of the IPCC. Knowledge held by state actors and environmental activists, as well as scientists, concerning climate change derives from IPCC activities and networks. International political attention to the possibilities of unmanaged climate change – the submersion of island states, loss of substantial areas of arable land and agricultural capacity – has been creatively institutionalized through the IPCC. Management strategies that disadvantage particular states or groups of states will not prevent unmanaged climate change. Such strategies will rapidly unravel as states ignore targets or opt out of international agreements.

Comprehensive negotiated regimes provide the only viable mechanism for international implementation of minimization strategies. The processes of international negotiation by which they are established necessitate relative fairness. In addition, they have an institutional capacity for integrating competing needs and interests. Comprehensive regimes entail sharing costs between parties, identifying targets for collective action and retaining common responsibility for implementation, including review.

Within the IPCC, scientific knowledge is a tool or platform upon which political notions and forms of organization can be constructed. In this way, knowledge becomes central to ideas of equity, responsibility, alleviation of climate change impacts and sustainable development. Knowledge is what provides motivation for moral action and strategies for ethical decision-making between states. In addition, knowledge underpins all developments in international law, but especially those concerned with increasing application of compliance or verification procedures associated with multilateral collective action agreements. Negotiated agreements encourage continued (post-agreement) interaction between actors, thereby enhancing flexibility within such agreements, providing opportunities for implementation review and renegotiated targets. Shared knowledge, contingent shared vision and policy options of scientific networks (epistemic communities) and state actors are vital in climate change management, especially the creation of institutionalized, comprehensive, climate change regimes.

The IPCC is particularly well placed to promote effective climate change regimes because it incorporates scientific expertise and state interests within its structures. It is a visionary body, seeking to find workable solutions to climate change consequences in which the burdens of strategies are shared in response to uneven socio-economic impacts as well as geophysical accidents. The IPCC is an extraordinary institution because of its capacity to establish relatively fixed bargaining coalitions across multiple interest sectors and because it recognizes that in negotiations towards climate change agreements and their implementation, knowledge is not apolitical or policy neutral. Rather, the IPCC proceeds on the basis that knowledge is a political instrument. This is particularly the case in relation to increasing overlap between the IPCC and FCCC, their shared communication networks, data bases and policy advisory mechanisms.

The IPCC, because of its links with other international institutions, cross-sectoral structures and multilateral membership networks, enjoys a privileged position in developing climate change regimes. Other climate change bodies remain reliant upon the IPCC for knowledge, the identification and redefinition of problems and predictions of consequences. The merging of scientific and policy-making interests and activities within the IPCC equip it for specialized management for effective implementation of climate change regimes. The IPCC is likely to expand its functions towards implementation review as climate change regimes develop more fully during the next cycle of negotiation, research, reporting and agreement formation.

Notes

1. Launched by the WMO in 1963, WWW involves the weather reporting systems of more than 160 states and is a key component of the Global Environmental Monitoring System (GEMS) coordinated by UNEP.

2. Epistemic communities are established international networks of experts who share a common knowledge base, common causal beliefs and a common worldview. Epistemic communities play crucial roles in three aspects of policy innovation: influencing the range and scope of political debate; contributing to definitions of state interests; and setting standards in information and choice alternatives.

4
The International Politics of Declining Forests

Minna Jokela

Introduction

Can forest resources be managed in a more sustainable way? What has the global dialogue on forests contributed to our understanding of the nature of the problem at hand? Environmental problems have become one of the most serious issues of international politics. The demand for governance in world affairs has never been greater. The actions of environmental movements and the subsequent increase in environmental awareness have exposed the need to pursue environmental issues at an international level, and one of the most important of these issues is the declining forests. Since the first observations of deforestation, from the end of the nineteenth century until the end of the Second World War, the problem was confined to the temperate latitudes. From the end of the Second World War, until the Earth Summit (United Nations Conference on Environment and Development (UNCED), 1992) the balance of the problem and hence the debate shifted to the tropics. The Earth Summit exposed the dangers of the irreplaceable loss of biodiversity, and since then the topic of debate has been expanded again to include the temperate latitudes.

We have entered a period of profound challenges to humankind's capacity to solve global problems. Massive deforestation has a high profile and therefore gains public attention. Several attempts have been made to create an international institutional framework for the management and conservation of forests. Although the Earth Summit failed to produce a forest convention, it did produce a statement on forest principles. Forest issues also overlap into the biodiversity and climate change conventions. The global forest debate is continuing but the pace is slow, as many unsettled issues burden the process.

Running parallel with the global process there are some important regional initiatives. Some international organizations like the European Union are ready to continue the forest protection process at a regional level, if global negotiations prove too difficult. The EU is today a key player in global politics. The EU is an interesting case to look at since it is, on the one hand, at the centre of the production of global risks and environmental hazards and, on the other, open to innovative ideas for ecological modernization. Lack of leadership hampers the global governance of forests even though global forest politics, during ten years of cooperation, have moved towards more responsible behaviour. The aim of this chapter is to examine the possibilities of ethically responsible leadership in global forest politics. I argue that the European Union is well placed to assert procedural leadership on forest policy in global fora because the institutional structure and politics of the EU itself provides a model of governance suited to late modernity.

As a starting point, I outline the evolution of global forest politics and evaluate the challenges that the global decline of forests poses to the international community. Then I discuss the role of the European Union in the modern world and note a number of features which show that the EU is becoming capable of taking the responsibility for the management of environmental problems in the international system. I conclude with a summary of the argument.

International forest politics

Forests cover approximately 30 per cent of the total land area of the world. They play an essential role both from the ecological and the economic point of view. Forests protect and stabilize soils and climate and provide the habitat for a vast array of life forms. Forests provide timber and also genetic material for industry, as well as reducing the level of carbon dioxide in the atmosphere, thus helping in the fight against global warming.

Forest protection is not a new issue on the agenda of various states. Biological conservation agreements are the oldest type of environmental treaties. The first were essentially bilateral agreements regarding the exploitation of single-species resources. Later the conservation of exploited wildlife became important. Recently, several multilateral treaties have been concluded to conserve wildlife habitats and species even though they were not being commercially exploited (Temple Lang, 1993). The internationalization of forest politics is a new phenomenon that has emerged since the 1960s when environmental problems began to be publicly debated.

The question of management and conservation of forests is complex. Present knowledge is limited, and decisions are made under conditions of uncertainty. Forests have many functions at different levels, and they are dynamic and always changing. It has, therefore, been too difficult to agree on how to manage forests globally. International forums on forests need to ask if 'management of nature' in the traditional, technical sense is at all possible, when we are living in a globalized 'risk society'.

The UNCED policy on forests

Apart from the 1983 International Tropical Timber Agreement, there has been no other international convention focused exclusively on forests. Before the Earth Summit, the international community possessed an inadequate legal regime to regulate the use and conservation of its forest heritage (Humphreys, 1996). Even though some developed countries expressed their desire to conclude a legally binding international convention on tropical forests, the forest issue was barely mentioned in the UN document that called for the UNCED. Nevertheless, forest policy became one of the most contentious issues.

The Food and Agriculture Organization (FAO) proposed a convention for the protection of forests. The tropical forest states, Brazil, Malaysia and Indonesia, opposed it on the grounds that they feared that an international code of conduct on forests, which considered forests as global commons instead of national resources, would require the imposition of unwelcome conservation policies. These countries formed a veto coalition to block the idea of a convention. The United States tried to achieve the constitution of an intergovernmental panel to discuss forests outside the UNCED framework, as was already the case for the issues of climate change and biodiversity. However the Group of Seventy-Seven developing nations (G77) refused to negotiate outside the UNCED framework. It was decided that the formulation of a forest convention would have been premature, and that UNCED could only agree in principle on a non-legally binding statement, which corresponded roughly to the will of the G77 nations (de Campos Mello, 1993; Porter and Brown, 1996).

Negotiations on forest issues quickly became polarized between the 'global responsibility' approach and the 'sovereign discretion' approach. For instance, the United States and Canada tried to link the principle of sovereignty of countries over their own forest resources with the principles of national responsibility and global concern for forests. On the other hand, for example, Malaysia and India saw these formulations as an effort to establish the legal principle that forests are 'global commons' or part of the 'common heritage of mankind' (de Campos Mello, 1993).

The forest question entered the global agenda in the late 1980s together with the biodiversity issue. At the UNCED they were recognized as global problems. The biodiversity convention and the forest principles are formulated around a broad ecosystem approach that acknowledges the complexity of ecosystems and the uncertainty of managing forest resources. It has been acknowledged that not all coordination of forest activities can be channelled through one process. Inevitably there will be various overlapping processes and institutions, both within the UN system and parallel to it, at both global and regional levels. This institutional diversity challenges the idea of convergence often required by international institutions.

Forest politics since UNCED

The squabbling between Northern and Southern timber-producing countries prevented a Rio convention on forests. A legally binding convention was not delivered, due to the suspicions of the South and the reluctance of the North to accept that they too were part of the problem. There was a lack of political will to admit that the protection of boreal virgin forests was just as important as saving the rainforests. Since the results were minimal the UNCED forest discussions gave rise to widespread disappointment and frustration among the participants. The failure to reach a consensus and the ensuing political debate led to a serious deadlock (Kolk, 1996).

From late 1993 onwards, a series of meetings was held in different forums. This development was facilitated by the decline of the North–South controversy on forests. Forest negotiations gradually began to focus on forest issues rather than political infighting over tangential issues. One factor accounting for this change of attitude was the recognition that the debate over tropical rainforests was not at all free from hypocrisy. The post-UNCED meetings increasingly encouraged the idea that it was equally important to preserve all types of forests. Other developments in this respect were the alarming reports about the consequences of global warming for boreal forests.

The post-Rio period, 1994–5, was one of confidence-building and emerging North–South partnerships, enabling the United Nations Commission on Sustainable Development, at its third session in 1995, to establish the Intergovernmental Panel on Forests (IPF), to continue the international forest policy dialogue. The IPF was established to promote the sustainable use of forests and to explore the possibilities of a global forest convention based on the forest principles accepted at Rio. The IPF met four times between 1995 and 1997, and the final report of the panel creates a central starting point and guidelines for international forest policy cooperation. In the report nearly 150 proposals for action were presented (Humphreys, 1998; International Institute for Sustainable Development, 1998; Commission on Sustainable Development, 1999).

The question of whether or not to start negotiations to promote a global forest convention was central in the panel's discussions. Since Rio there have been some changes in the positions of governments. During the UNCED forest negotiations the South was strongly against a convention, whereas the North argued in favour. Since Rio the North–South polarization has blurred. Forest convention is much more attractive to the South if Southern nations can perceive a relationship between forest conservation and financial and technological transfers. Most Latin American countries, however, still argue against a forest convention. The common policy line of the South is that it is still too early to start the formal treaty negotiations but the forest dialogue should continue (Humphreys, 1998).

At its nineteenth special session in 1997 the UN General Assembly decided to continue the intergovernmental policy dialogue on forests through the establishment of an Intergovernmental Forum on Forests (IFF). Its brief was to identify the possible elements of, and work towards, a consensus on international mechanisms such as a legally binding instrument. Although the global forest dialogue has brought the South and the North closer together, delegates were encumbered by the same political baggage. While delegates were aware of the need for progress, many issues which were unresolvable at IPF remained problematic (International Institute for Sustainable Development, 1998; Second Session of the Intergovernmental Forum on Forests, 1999).

The global forest policy debate is suffering from a lack of clear identity. Two series of intergovernmental negotiations (UNCED and IPF) have failed to produce a global forest convention, but the dialogue has continued. In each new phase of the process the same ideas re-emerge without a clear knowledge of the role of the new institution. It has been argued that this lack of identity follows from a lack of leadership. International organizations have been criticized for their inability to fill the leadership vacuum in international forest politics. Delegates have used the proverb 'too many cooks spoil the broth', and at least one delegate has argued that a 'head chef' is needed to address the gaps between existing instruments (World Resources Institute, 1999; Second

Session of the Intergovernmental Forum on Forests, 1998). Scientific communities played a vital role in biodiversity and climate change issues; they have been important in making the international community understand the importance of a global convention. The European Union has exercised influence but a clear leader in the debates has so far failed to emerge. Nevertheless the European Union may yet provide a model for environmental governance.

Europe in late modernity

The European Union is a unique institution purposely founded to solve the problems of late modernity characterized by widespread scepticism about rationality and knowledge. The transformation of the unseen side-effects of industrial production into global ecological trouble spots constitutes a far-reaching institutional crisis for industrial society itself (Beck, 1996; Elliott, 1995). This leads Beck to claim that present-day life is moving towards a distrust of expert knowledge (Beck, 1994: 28–31). Elliott argues that social institutions have become subject to reflexively organized domains of action, whereby thinking in terms of security and risks is fundamental to the reproduction of society. Reflexivity should be conceived as a continuous flow of individual and collective 'self-monitoring' in which room is made for the contingency and openness of social life. To live in the late modern world is not just to live in a world subject to more or less continuous and profound processes of change; it is also to live in a world where change does not consistently conform either to human expectation or to human control (Elliott, 1995).

The issue of reflexivity provokes the question: what is really the European response to the problems of late modernity? EU integration is a dynamic, open-ended process in which the end product is dependent on the path taken in the negotiations. The EU has been in iterative change for most of its history. The result is an extraordinarily dynamic polity in which innovations agreed by national governments in the major treaties create webs of constraints and inducements that lead to new conflicts and pressures on governments for further institutional negotiation (Marks et al., 1996).

How 'Europe' has come to be seen as a political space is a long-term process which has become woven into the fabric of European society. The EU and its institutions are based on more than just the 'political will' of heads of governments or legal treaty properties. They are reactions to persistent and fundamental patterns of West European political and social developments. The European project is driven by more than merely economic rationality. It is also, crucially, based upon those reflexive foundations that provide the legitimacy and purpose for common action (Christiansen, 1997).

The consequences of modernity (both positive and negative) can possibly be best seen in Europe. The EU is facing problems with what has come to be referred to as the democratic deficit or legitimacy crisis that is one part of the larger crisis facing modernity's institutions. Therefore the EU is today at a crossroads: it can remain an intergovernmental institution or it can become a supranational actor. Depending on which path the EU takes at this juncture, it has the potential to aggregate conflicting interests and to introduce new measures in international forest policy. Given its particular design, the EU may be more potent, efficient and robust a body than any other international institution (Kux, 1997). The intergovernmental nature of EU decision-making may clash with attempts by the EU to present itself as a supranational actor capable of taking the leadership role. However it has the potential to become both an important intermediate platform and a pioneer of international environmental cooperation. What we have in Europe today is a novel system of rule that brings into question our Westphalian vision of what international politics is about (Wind, 1997).

Liberatore makes several points about the role of the EU as a representative of its member states and as a distinct actor trying to play an environmental leadership role. When different views and interests of the member states are mediated and agreement is reached prior to international negotiations, the EU can speak with one voice and advocate positions and solutions that are not simply the lowest common denominator of diverging national positions. Unfortunately, the time necessary to reach such internal agreement rarely coincides with developments in international environmental negotiations. In addition, the EU has to deal with complex legal provisions regulating its external responsibilities. Because it does not have exclusive authority in the environmental field, the EU can only be a party to international conventions in the form of 'mixed agreements' where both the EU and its member states are parties. Under such mixed agreements, the Commission can negotiate on behalf of the EU only if granted a unanimous mandate by the Council (Liberatore, 1997).

The evolution of the external role of the European Union is a multidimensional process. It includes, for example, the building of formal rules and institutions as well as the development of informal practices and action against third parties. One aspect of the dynamics of the actor capability of the EU is the formation of a collective outlook pertaining to the Union. Global pressures and other external influences demand that the EU acts in harmony. Coordination at the global level leads the Union to build up actor capability. Internal integration and external integration interact. Effective external policies require the EU to adopt internal sectoral policies to support global politics. External influences, especially pressures rising from global problems, explain why actor capability comes to be created within the Union.

The forest issue and the European Union

The need for EU leadership on forest policy

Although the global policy process has been hampered by lack of clear leadership, the EU has always been a key player. One of the most important factors in the development of EU forest policy has been the appearance of environmental questions on the agenda of international politics as well as within the internal politics of national societies. In industrialized societies public opinion has been an instrument in demanding active forest protection measures and this has been reinforced by the action of citizens in their role as consumers. As the EU's role is important in the global forest debate, cooperation in EU forest politics becomes essential, as the credibility of the Union in global negotiations requires concrete actions at a regional level.

Forest policy in the European Union has never been one of the main areas of integration, and for this reason this policy domain has attracted little scholarly attention among the analysts of the EU. However, since the late 1980s forest politics has become one of the concerns of the many actors in the EU who give rise to the various initiatives. Forest policy is also one of the fields in which there have been major and persistent conflicts of interests between the Commission, member governments and interest groups. The role of member governments in this policy domain has traditionally been dominant, and in many ways still is. However, the emergence of the EU as a major actor in European forest policy in the 1990s has brought about a decisive shift in the importance of forest policy and its place on the EU agenda compared with the preceding period.

Although the EU has no common forest policy its Commission has paid growing attention to forest issues and in 1989 the EU adopted a Community Action Programme (Commission of the European Communities, 1988). The EU has also acted strongly at international forums on forest protection. In the Rio Conference the EU acted as a unified group on many forest issues. It was also an active participant in the Ministerial Conference on the Protection of Forests in Europe and a signatory of the resultant resolutions (Ministerial Conference on the Protection of Forests in Europe, 1995).

In the 1990s the forest debate entered the European Parliament. It is a contentious issue in the Parliament, and it is interesting to look at the problem definitions and justifications in the European Parliament documents (European Parliament, 1994, 1996). For several years now forest politics has had a global dimension. Global problems demand that the EU work more effectively on forest issues, since the Union cannot be a credible and effective actor in global forums if it does not show itself to be a good example to the rest of the world. Stronger actor capability provides the Union with a greater possibility to speak with one voice and deliver a clear message at global forums. Since forest issues have moved to the forefront of the international agenda, social and environmental functions of the forests and the internationalization of the forestry debate have created new challenges.

There are basically five reasons why the European Union needs a forest policy. First, widening of the Union to the Nordic countries brings in large forested regions. (Membership of Finland and Sweden doubled the forest area of the EU. With Austria they made the Union self-sufficient in timber products and the importance of forest issues grew.) Second, and similarly, widening of the Union to include the Eastern European countries includes new forests. Third, the public's interest in and expectations of forest policy has grown in recent years. Fourth, the confusion over the governance of forest issues in the Union stimulates debate. Finally the Union is increasing its role in global forest negotiations.

The European Union is confronted with the challenge of forging a strong integrated policy in keeping with the national policies pursued by its member states in regard to tropical forests, and the adoption of a clear stance towards new member states. The forest is perceived as an element of humanity's collective heritage, similar to air and water, to which everyone has rights, and for which no one can be held individually responsible or accountable. Whether it is a matter of the individual heritage of its member states or of its intervention in the interests of the tropical forests, the European Union is confronted with crucial questions in the global debate. Ecological, economic and social stakes extend far beyond the scientific and technical capabilities of any single nation and the forestry measures introduced at EU level are overcautious. Therefore, without a common forest policy the EU is only making itself heard with a muted voice within international circles.

For decades the EU has failed to see forest politics as an issue requiring common policy. However, the global problems and developments in global policy coordination have revealed the need to speak with one voice in the global arena. External pressures and influences have been a driving force towards the coordination of policies within the Union and it has now reached a critical phase in its evolution. It is faced with new demands: the social and environmental functions of forests and the internationalization of the forestry debate, and in these areas it is being subjected to new and contradictory pressures.

Procedural leadership: a broker role in global politics

Successful global politics requires clear leadership and some actors who actively pursue the interests of all. The traditional approach of self-interested states, maximizing their national interests is, however, unable to explain the development of global policy. Kaelberer writes that a leader has to perform the function of a 'broker'. Cooperation does not happen automatically but needs to be created. The leader has to bridge the various distributional concerns associated with cooperation and forge consensus amongst potential cooperators. The broker is one of the interested parties in the negotiations. The broker brings the involved parties to the bargaining table and establishes the agenda. When necessary the leader must solicit concessions from the participants in order to establish a compromise that all parties can accept as a gain over the status quo. If distributional concerns on the part of some or all member states characterize the policy realm, leaders have to find areas of agreement that satisfy all parties. Most crucially, the broker itself may have to make concessions in order to achieve a solution – an act that distinguishes this kind of leadership clearly from an ordinary third party broker (Kaelberer, 1997).

This broker role is not hegemonic or completely dominant. Appropriate leadership is required if established institutions are to develop to the point where successful cooperation is possible; however, deprived of this leadership, institutions become fragile and cooperation becomes difficult if not impossible. By the early 1990s a few states had begun to regard the role of leadership on the global environment as a means of enhancing their international status: both the United States and Germany made bids for leadership of a possible forest convention in 1990–91. The European Union also had aspirations for leadership on climate change politics and in the Rio Conference asserted that role in order to signify its emergence as a global power (Porter and Brown, 1996).

The EU can play a useful role by facilitating the evolution of global environmental policy. But it also suggests an institutional model of governance because of its particular structures, rules and procedures. The EU has developed a new form of politics, involving a quality and intensity of cooperation at all levels, across multiple levels of government and between private and public actors, promoting social learning. A grid of dispersed and often fragmented political domains facilitates a discourse on ecological modernization and risk management, around which policy could be organized and discussed. Kux argues that the institutional setting of the EU is well suited to deal with current and future environmental conflicts. It has the potential to aggregate conflicting interests and to introduce new measures in a gradual, negotiated way (Kux, 1997). Clearly the EU has the potential to gain the procedural leadership in global forest policy.

Conclusion

The evolution of global forest policy began in the 1980s. At the UNCED the forest issue became politicized and drifted into deadlock. However, the forest debate had started and many actors have since softened their attitude on the forest question. Recently, the EU has faced problems that stem from deeper processes of modernization and industrialization. It has been acknowledged that the forest debate has moved to the forefront of international politics, and that the decline of forests is a global problem. In this field it has been recognized that uncertainty exists, since ecological stakes extend far beyond the current scientific understanding and capabilities. External pressures and influences have been recognized as a driving force for mutual cooperation. Global problems have revealed a need to speak with one voice. It is important for the EU that European identity become stronger in relation to forest issues.

The EU is confronted with crucial questions and faced with new demands in the global forest debate. Due to the contradictory pressures present in that debate, leadership that differs from its traditional hegemonic or dominant style is required. The EU has emphasized that the need for leadership is important within the global political arena. In the EU documents the need for concessions is acknowledged: the EU cannot be considered a credible and effective actor in global politics unless it leads by example. In the field of forest politics it still has a long way to go. The EU's institutional structure is open to new ideas, and therefore it has the potential to aggregate conflicting interests and to go further in fulfilling the broker's role within the arena of global forest politics.

5
Maximizing Justice for Environmental Refugees: A Transnational Institution on Behalf of the Deterritorialized

Adrianna Semmens

> *There is one and only one social responsibility of business – to use its resources in activities designed to increase its profits* so long as it stays within the rules of the game [My emphasis].
>
> (*Friedman, cited by Cadbury, 1998*)

Introduction

The issue of the migration of people whom I will term 'environmental refugees' raises serious questions about our current normative frameworks of justice and governance. Questions about the nature and extent of our responsibilities towards migrating human populations take on a new meaning when many of these migrations are 'forced'; induced by the adverse environmental effects of development policies. More problematic still are questions of responsibility when transposed as policy dilemmas such as the following: What would resettlement mean under increasing widespread conditions of environmental degradation? Considering it would be folly to suggest the siphoning of populations from one area of environmental decline to another, say Shenyang to Shanghai, where would one transfer or siphon people? How could one weigh the rights of environmental refugees to migrate following displacements, against the rights of local populations and/or the non-human world? Who would finance social welfare programmes and safeguard human rights?

It is too soon to say whether such normative policy dilemmas will be entertaining non-Western and Western policy sophistries in some future time, and/or whether such dilemmas will be abandoned for authoritarian quick-fix solutions, if the phenomenon of environmental degradation and environmental refugees continues to increase. Either way, we, non-Western together with Western earthlings, will be increasingly lowering our earthy standards of existence. Is that what you and I want?

What I explore in this chapter is how international interventions can be topologized, in terms of justice and governance, to redress the plights of environmental refugees in the present, to minimize further injustices to diverse lands and diverse peoples. Can the plights of environmental refugees be redressed through the creation of an *international environmental refugee* regime?

The current international refugee regime set up to address the needs of traditional 'political' refugees in sudden onset emergency situations will not do. Already this Eurocentric regime (Tuit, 1996) is overburdened and ineffective, its justice reduced to a matter of western restrictivist insurance stances and to intergovernmental burden-sharing under an ambivalent comprehensive security discourse (Hathaway, 1997). To date, there are no legitimate authority and/or regularized patterns of cooperation, with the mandate to act in the interests of people directly and indirectly displaced as a result of development-induced environmental degradations.

By way of situating my discussion in contemporary discourses, I begin by introducing the issue of the migration of environmental refugees and problematizing its androcentric geopolitical security conceptualizations. I follow this with a discussion of the plight of environmental refugees which brings to light the need for a notion of justice which takes cognizance of interlocking oppressions. My discussion draws attention to those interstices in the political economy and politics of many environmental refugee migrations where action is often overlooked by policymakers. Relevant examples include: the sourcing and collaborations of transnationalism; the proletarianization of rural people; the feminization of labour; the constructions of the 'alien other'; and the deprivation of human rights. In the final section, I reflect in a preliminary way upon justice and its institutional corporeality for environmental refugees.

Environmental refugees: a homeless issue in scholarly canons

For decades state urban territorialization practices in concert with elite transnational corporate leagues and multilateral development projects, have despoiled ecosystems and contributed to land-watershed degradations. Through activities such as agribusiness, structural adjust-ment and infrastructural development programmes they have coercively uprooted from their social-environmental domiciles – that is, deterritori-alized – millions of poor rural families and communities (Gadgil and Guha, 1995; McCully, 1996).

Now, under the 'transnationalization of the world economy' (York, 1988: 1) the phenomenon and sourcing of deterritorialization is increas-ing. Across the world ex-subsistence rural and urban folk, particularly women and their families, are compelled to choose the livelihood strategies of moving within a state and/or across state borders because of development-induced environmental degradations. Some such migrations make Western media headlines, as in the case of people flee-ing official ecological disaster zones in Russia like Tagil (Yablokov et al., 1993). However, the migration of people indirectly impacted by adverse environmental consequences of development policies goes unrecog-nized (Pearce, 1991).

This raises the issue of how to refer to this group of largely unrecog-nized people. I refer to them in this chapter as 'environmental refu-gees'. However, given that any categorizations can be thought of, in the Foucauldian sense, as dominations linked with the treatment of those categorized in punitive ways, is it politically prudent to use the category 'environmental refugees'?

Moreover, the use of the term 'environmental refugees' can be con-tested in other ways. For instance, one could problematize the use of the term 'refugees' and the extent to which the elements of external force, and targeted persecution (those baseline conditions that socio-logically and legally help to define refugee status) actually apply to *all* people *indirectly* impacted by development-induced environmental degradations. For example, ought a thirteen-year-old Thai girl, who leaves her home in the province of Chiang Rai for the city of Pattaya to sell sex, qualify as a 'refugee'? Arguably, a conceptual 'line' (that rule purged of emotion) between environmental migrants and environmen-tal refugees ought to be drawn. But where? Let us hear her speak openly: 'The village is no longer a safe place for us. Nor is our fam-ily ... ' (Wiyaprao, 1996). It comes to light that most of her village had been taken over by land speculators and that her parents are now bankrupt. What if such circumstances were suppressed from her view, from our view?

One could also call into question the identity which the official term 'refugee' constructs in the late twentieth century (brown-skinned, gender-neutral, 'alien'). More critically still, one could challenge the basis of our sociological and legal understandings of the politically malleable concept 'refugee' and reconstruct its meaning by critically scrutinizing the theoretical salience of 'new' (some progressive, some conservative) ecological theories and feminisms. There is a need to examine thoroughly all these problematizations by engaging with a range of dialogues, but such a task is beyond the scope of this chapter.

In this text I reserve the term 'environmental refugee' for people who have been grossly discriminated against, persecuted, impoverished, and who have a basic need for food and safe shelter, as well as a need for global citizenship, integrity, fellowship and creativity. I suggest that deterritorialization (and its inscription) for such people signifies something markedly different from the sort of deterritorialization condition experienced by many androcentric, disembedded members of transnational business consortia. (These can include, bankers, government officials, and military personnel.)

To be deterritorialized by the slow onset of adverse environmental effects of development policies and projects is not experienced as a single event. People often are terrorized physically as well as verbally, and/or made insecure, long before they become finally deterritorialized from the social and environmental webs that have hitherto sustained them. To a small land farmer or *minifundista* (an El Salvadorian term for a female smallholder farmer), the terrorization can begin with the end of a gun or other tools of male machismo. For another inhabitant of Loa Tepu, a village in Indonesia, it can begin with an exploration team from a transnational extractive company and an official letter. For a coloured woman living near Darwin, Australia, it may come with a direct, patronizing gaze from a politician with an environmental portfolio concealing a cartography of uranium mines.

By contrast, I propose that 'deterritorialization' in the case of androcentric transnational consortia, means to fall prey to the collective pathology of the 'escalator fallacy', to borrow Midgley's metaphor (Midgley, 1985). This means succumbing to an unexamined, arrogant faith in a distinct form of human evolution that holds the remote, greedy Faustian man/woman as representing an intermediate stage in a line from primates to a higher form of human possessing enormous powers to subjugate nature. Under this evolutionary quest, some material bodies and social, ecological webs are mythed as inferior forms which can be plundered and exploited. Fox (1990) would rightly term this view 'phallicism', a worship of 'up-ness' and 'high' living. Following the many self-defined feminist environmental philosophers and ecofeminists, I suggest that under this fallacy our ecological roots become associated with the domain of the female-bodied, the realm of animality and the realm of empathy and sentiments which one must take flight from (Plumwood, 1993; Warren, 1994; Salleh, 1997).

In the past, some northern environmentalists, among them funda-mentalist deep ecologists like Abbey, approached the issue of the migration of environmental refugees as a 'question of people [from the developing worlds] moving beyond the carrying capacity of the land' to 'developed' countries (Weinberg, 1991: 163). As for conventional state functionaries, they have downplayed the issue, framing many contemporary environmental refugee migrations as economic ones and categorizing people under a range of diverse categories, such as squat-ters and/or illegals. This effectively places groups of people, indirectly affected by development-induced outmigrations, off the socio-economic political map. Consider one woman's testimony:

> We are called squatters. How can we be squatters in our own country. If we are squatters, what are all the foreigners who come to buy up prime land? ... The truth is we are refugees in our own land. In Bicol, we were happy to grow rice, bananas, coconuts but we cannot plant now. Everything that is planted is just blown away by the wind.
>
> (in Seabrook, 1993: 39)

In recent times, the issue of the migration of environmental refugees in contemporary scholarship had been largely homeless, exiled from inter-national refugee policy analyses as well as from androcentric environ-mental policy and regime discourses. While loss of biodiversity, ozone depletion, and a host of other environmental problems crowded the envi-ronmentalist frame, the issue of environmental refugee migration roamed in the netherworlds of global environmental agendas. Now finally shack-led, it is largely confined to the turfs and fiefdoms of ambiguous, subtle neo-Malthusian, geopolitical-ecological security discourses (e.g. Myers, 1986, 1995; Ophuls, 1992; Richmond, 1994; Black, 1998).

But this issue does not rest easy in these androcentric confines. Why do studies downplay the maldevelopment and political mismanage-ment dimension? Where are the concerns of environmental refugees, particularly 'women', to be found in this canon? Why do androcentric analyses take cognizance of the most dramatic direct cases of environ-mental outmigrations from the 'Third World'? Is it possible for envi-ronmental refugees to become illegal and economic migrants? When do environmental refugees lose their 'refugeehood'? Why does the issue predominantly fall within the ambit of 'life-boat ethics'?[1]

In light of this, the issue of the migration of environmental refugees needs to move beyond the stranglehold of the population–security nexus. To suggest this, does not mean, of course, that we ought to deny the adverse environmental impacts of large human population movements and/or geopolitical issues of asylum. However, I would contend that the population–security nexus narrowly circumscribes not only the issue of the migration of environmental refugees, but along with it the issue of justice and its institutionalization. We need to enter empathetically into the worlds of environmental refugees before any discussion of justice is possible.

Travelling to environmental refugee migrations[2]

Could we bear witness to the complex ambiguous social realities and perils of heterogeneous forms of development-induced environmental human displacement in the limited space of this text? Can our mental worlds, situated as they are in cultural, racial, gendered, ecological and historical assumptions, effectively characterize the plights of European as well as non-European people who become environmental refugees, without *gross* misrepresentation?

To minimize what Haraway (1991) refers to as the 'god trick' – the view from nowhere[3] – I suggest we begin by adopting the subject posture of the border crosser intellectual[4] which welcomes diversity and acknowledges voices exiled from mainstream disciplinary discourses. Important theoretical and practical insights can be drawn from a melange of multidiscursive genres, notably political ecology, subaltern studies, postcolonial studies, ecofeminism, critical development studies, postmodern and materialist feminism.[5]

The following is a guided tour into and across the world of environmental refugees. This tour does not purport to be comprehensive. Its purpose is to provide a general picture of the migrations of environmental refugees. Needless to say, the homologies which I present in the tour serve as mediums for compressed information; the pattern of relationships that I construct does not always appear in this form in the lived experiences of environmental refugees.

The tour borrows from the artistic sensitivity of the 'world travelling approach' as notable feminists like Lugones and Sylvester describe it.[6] In Sylvester's words, (1995: 946) 'This form of world travelling relies on empathy to enter into the spirit of difference and find in it an echo of oneself as other than the way one seems to be.' The tour is both ironical (in the sense that situated panopticism can lapse into totalizations)

and full of ironies. In part, some of these ironies are due to the contradictions that may arise from our own shifting identities in modern
society, while others are due to the hypocrisy and contradictions of
privilege.

As you enter the world of the environmental refugee, you experience
niches and alleyways that have been obscured, criminalized, marginal-
ized and demonized in traditional political geographic mappings of
terrestrial space. You[7] enter a world of farm invasions, marches to mar-
ginal lands or forested areas, arduous trekking from squalid urban shel-
ters and shanty towns; from detention centres and refugee camps; back
and forth from highland village to coastal village, airports and leaking
boats. This world is diverse yet also janus-faced. It is a place inhabited
by the resourceful and those that empathize with them, as well as a
place which the unscrupulous deterritorialized entrepreneurial elite
and their apologists traverse.

> You usurped my grandfather's vineyards
> & the plot of land I used to plow,
> I and all my children,
> and you left us ... nothing but these rocks. ...
> I do not hate people
> However
> if I am hungry
> I will eat the flesh of my usurper;
> beware, beware of my hunger and of my anger.
>
> (Darwish, 1996: 11)

[The underlying] principle of aid ... is that you give away a little to
make sure that you can keep a lot more ... First you transfer billions
of pounds from them, then out of a sense of responsibility you give
part of your billions away in charitable acts.[8]

In this world, you, the environmental refugee, will barely survive. But
considering that the concept 'environmental refugee' has little currency in the minds of mainstream scholars and officials this will not
unduly immobilize you. (Unless of course you die first.)

While some of these scholars refer to your migration as adventurous
and economically motivated, necessity, not adventure, bright lights
and profit can often guide your movements (Seabrook, 1993). Some of
you however may have been so severely stung by European civilizing
missions and their cargo of cheap goods that these will lure you with

promises of Eden. This means that you will ultimately be pushed to move by ecological decline on the one hand, and lured by the gold of progress prophets on the other. Political officials from the halls of political and economic power will ignore the push factor and focus on the lure factor. They will expect you to go home even after the territories you were compelled to leave possess little capacity to reintegrate you.

When your land in a Russian province has been polluted by heptyl rocket fuel, or you have been uprooted because of agricultural progress in China, or maybe even through the activities of multilateral agencies like the International Monetary Fund together with the World Bank...

> Tell me, how do you go home? ... They built dams in Rajasthan too. There too they promised all sorts of things ... land for land, compensation in thousands, everything. The people ... gave up everything ... Do you know where they are now? Scrounging around in garbage heaps of big cities hoping to snatch a morsel of food.
>
> (Swaminadhan, cited in Rich, 1994)

Given the opportunity, you will wonder about the logic of bilateral and multilateral aid programmes since you may have been displaced, first, for poverty alleviation projects like commercial logging and mining, giant hydroschemes, agribusiness, golf courses, cheap food imports (and convention complexes attended by World Bank and IMF delegates); and then for loan rescheduling programmes which often encourage the same thing.

To make matters worse, many state functionaries, some of whom may belong to exclusive golf clubs, will strip you of your personhood. This means that you will be defined as part of a *flood*, or *flow* of refugees. Unofficially, you will be regarded as a non-person. You! The proud Yanomani Indian from Brazil, the Asmat native of West Papua, the Hindu Indian, or even the European ex-subsistence farmer will be regarded as a barbarian (Cela, 1996)!

As a barbarian, many western and non-western transnational logging and mining companies will blatantly ignore your human and legal rights as they destroy your forests, forcing you to labour for a little shag tobacco and cheap clothing (Colchester, 1996). If you happen to be a female, your economic dependence on male members of the family may increase along with your social responsibilities (Atkinson, 1998). Yet persons like you will be attributed with immense powers of destruction over the earth's physical environment (Kolle, cited by Gan, 1993); powers greater than the combined forces of the growth development paradigm, international criminal organizations in concert with some bank consortiums, government sponsored transmigration programmes or even development finance institutions.

Given your perceived powers, you are unlikely to qualify (or have the money) for legal cross-country migration. You may prefer instead to migrate within your own country. As an internally displaced person, don't expect to qualify for Convention refugee status. Therefore, typically (unless your situation is defined as a complex emergency arising from war or 'natural' calamities), you will not be eligible for international protection, food aid relief, or other assistance.

But what if by some stroke of fate, some say ill fate, you managed to migrate alive (through illicit trafficking operations) to another country on some rickety boat? Then as a 'illegal alien', your human rights will be compromised under state sovereignty rights and the stranglehold of detention centres designed to deter others like you from entering the country (Bosniak, 1994).

Weren't you aware by now that as part of a 'flood' or 'flow' of millions of 'illegal aliens' worldwide, you need to be contained? Never mind the fact that cosmopolitan transnational shareholders (some cooperating with large-scale criminal syndicates (Chossudovsky, 1997) or even those who purport to be concerned with your welfare, are at this moment boarding their planes for Europe after lunching in their executive dining rooms!

> [For the record, at a recent IMF–World Bank annual conference no expense was spared] Guests began with crab cakes, caviare and … mini beef wellingtons. The fish course was lobster … the entree was duck with lime sauce.
>
> (Hancock in Korten, 1995: 104)

As an 'illegal foreign alien' or more euphemistically termed 'undocumented migrant', don't expect to qualify for Convention refugee status unless you can prove to be a bona fide refugee. This means you have been fleeing direct political persecution or severe political instability, or have experienced a massive violation of human rights. But even this will not guarantee you political asylum. Remember what happened to your so-called political refugee neighbours from Haiti when they sought asylum in the United States? Ultimately they suffered in detention centres and were repatriated (Abbott, 1988), so you might as well forget your well-rehearsed line 'I want political shelter'. It does not always work. But asylum is the least of your problems.

If you, the 'alien', happened to have emigrated to another political territory (say from Bangladesh to Italy or Spain because you were fortunate to belong to the educated middle classes), your passage to your final destination could well be an indirect one via a series of journeys, fraught with perils. Not all of you will survive of course.

> My name is Genaro-Cux-Garcia. I am speaking to you via an interpreter. I have survived the truck accident which killed my fellows. It has been thirty days now since I left my Ranchito in rural Guatemala. My wife and two daughters wait for money to pay for corn and beans. I have nothing to send them. Nothing. I could lose my tiny plot of land because of the deed I signed when I borrowed money. My wife, she said 'Genaro don't go' what if we can't repay the money? ... The $US 2000 I borrowed is gone, some I gave to bandits, some I gave for bribes to police and some to the truck driver.
>
> (adapted from Borgman, 1996)

Meanwhile,

> Men ... cursed they should be because they live on what their women do. The women go to the field, to the vineyard. They harvest and do all kinds of work day and night.
>
> (Cajupi in Pritchett-Post, 1998: 14)

Those of you who survive will require employment or some form of welfare payment. Consider yourself fortunate if you can survive on hawking or maybe even on some dynamite fishing on a remote Pacific island. Jump for joy, if you qualify as a low-wage supplier in the low-wage market for US Japanese and West European Transnational Corporate interests (Pettman, 1996).

So you managed to obtain a job as a field worker in an export-oriented Chilean fruit industry? The one which imports banned pesticides from the United States controlled by US companies? 'Please don't look at me that way! ... I know my baby was born without a brain, but other women have babies like that!' (adapted from Casagrande, 1996).

Perhaps you or maybe your children have obtained employment working in the sweatshops of the global factory making electrical goods, or clothes, 'toiling 60 hours a week' and making US$30 a month as rats and mice crawl over your feet?

Look don't point the finger at us retailers. We don't employ these workers.We are removed at least two or three steps from the problem. Point the finger at our sweatshop subcontractors!

(adapted from Branigan, 1997)

You may be fired for yawning in one of the many sweatshops, some owned by strictly economic migrants (although from your experience you know that the line between economic and environmental can often be blurred).

O.K. so I fired her! She wasn't quick enough. But I'm at the bottom of the garment industry. The big-name stores and fashion labels, they don't know the troubles they cause me. The garment manufacturers who subcontract work to me owe me thousands of dollars. I've threatened legal action.

(adapted from Branigan, 1997)

Perhaps you have left the vestiges of morality behind and became involved in pernicious ventures such as trafficking in human lives (United Nations High Commissioner for Refugees, 1995), animal hides, and/or piracy?

Whatever you do you will be regarded as a burden even though many transnational accumulations of capital are dependent on your labour. Your topics of conversation will include your wretched poverty and how to alleviate it as wealthy shareholders wing their way between offshore banking havens like the Caribbean islands perhaps engrossed in the same thing. Consider this fanciful dialogue in a Socratic vein:

Tom: Look, Socs, don't make it a philosophical argument. We need the sword! Coercion is not necessarily a vice. How else are we able to restrain the illegals? They're arriving from all over the place. Now I hear they're gaining entry by posing as foreign businessmen on intracompany transfers. The cheek! Sure we need coercion. We've got to have order don't we? Surely coercion is vital if we want some form of economic prosperity (adapted from Branigan, 1996).

Socs: It is a philosophical argument as far as I'm concerned. I think the world needs a little less coercion; (winking) a light hand in the till and a lighter hand on the tiller.

(Clicking his fingers) Hostess, get that baby in economy class to shut up, and bring me some coffee!

If you happen to be female as well as an environmental refugee, you may be versatile, resourceful, intelligent, but you will probably have additional problems to contend with.

He orders coffee. What does he know? I lost my brother in the coffee plantation. He was sprayed by plane while we were working and died soon afterwards. I was ordered to leave. I am grateful to have this casual job for these airlines. My companera in India was unlucky. Like me, she fought for secure rights to her land but when she became widowed it was her mother in law who bought her a one way ticket to Varanasi.

(adapted from Durning, 1989)

Places such as Varanasi, known for almost a hundred and seventy years as a refuge for widowed Hindu women, will not always welcome you.

All is not well here. I've been treated like a leper … but I hope that Conchita's fate was kinder. After her husband abandoned her when she was fired from the coffee plantation, she took her children to live in one of the favelas in Sao Paulo.

(adapted from Seabrook, 1993)

You will probably feel unsafe in squatter settlements and pass the night in terror, particularly if you are widowed or a single parent. You will have little privacy where you live because your tarpaulin makeshift shelter is unlikely to have secure doors. You pray for menopause, and then there is the crying of the children.

You may become vulnerable to further gender-based violence, before, during and after displacement and may fall prey to unscrupulous smugglers who traffic in women.

Don't call me a poor thing. After he told me to leave the plantation because I complained about the poisons, I met this woman from Chile. She said she knew a man called Juan who could find me a job in Holland as a dancer. He took me to the Belgium embassy and gave me tickets and money. When I got out of the embassy he took them back. It was show-money. He said he would lend me money to go to Holland. When I got there, I thought 'Oh God, I can't do this' but he put a knife to my throat. The money I make now I send to my parents back home. I will never forget one of my customers.

(adapted from Altink, 1995)

One of your customers may be an executive of a transnational corporation.

Heck ... If I knew she was with me against her will, I would have left immediately! I'm not that kind of man! I'm an executive for a transnational company, but listen here I'm not in the business of exploiting women. We all do it one time or other, visit the shows. I was after some fun that's all ... you know what I mean. Point the finger at Juan the brothel owner, not me, for her fate!

(adapted from Enloe, 1990)

Having been deterritorialized from your land and cultural roots, having borne disproportionately the social costs of displacement and survived, you may not want to go back home.

[O]nce home was a place, perhaps the only place where I imagined that I really did belong where I thought myself whole ... This is not so anymore ... The word 'home' and all it represents has shifted its meaning too many times.

(Danquah, 1998)

Just-talk or maximizing justice for environmental refugees?

[A] reduction in the Cadillacs of the few, could I believe, provide bicycles for the many.

(Hathaway, 1997)

Clearly there are a range of interlocking struggles in the face of interlocking oppressions that often mark the plights of many 'environmental refugees'. What becomes apparent from the foregoing discussion is that working within the bounds of international distributive justice (e.g. financial, resource, and 'burden' distributions) would narrowly circumscribe justice and its interventions. Notably, financial and resource distributions have in the past become transposed as bilateral and multilateral aid programmes which have deterritorialized people.

Also people gendered 'women' in remote villages and/or people in detention centres and rickety boats are unlikely to see international distributive justices. What would international distributive justice mean anyway without human rights, compassion and the provision of non-exploitative livelihoods?

Suppose enlightened environmental non-western and western activists, scholars, and policy-makers wanted to enlarge the notion of justice as well as to increase its likelihood for environmental refugees. How would the formation of an institution, namely, an international environmental refugee regime, fare? This depends in whose interests such a regime, is created and implemented for, and for what collective outcomes.

Briefly, there is little doubt among many western neoliberal environmental institutionalists that environmental regimes are needed to tackle contemporary cross-boundary environmental issues. Environmental regimes, as the argument goes, matter because they make cooperation possible on collective problems by the process of collective decision-making. The view seems to be that any regime is in some cases better than no regime at all especially since it exerts an influence on some forms of collective interactive behaviour.

But environmental regimes are partial normative systems of governance whose formation emphasizes certain values, structural norms and implementation actions over others. Young (1994: 3) has suggested that regimes are essentially institutionalized norms, i.e. 'sets of rules of the game or codes of conduct that serve to define social practices, assign roles to the participants in these practices, and guide the interactions among occupants of these roles'.

So what are these 'rules of the game' or codes of conduct embodied by the formation of an international environmental regime? Regime frameworks of governance typically delimit governance to a given area of international relations, so that the 'regimentation' of an environmental issue in practice is likely to reduce the issue to an international problem. More precisely, this means that the normative system of the international society of states, notably geopolitical notions of space as state resource containers, along with 'foreign' policy decision-making and issues like sovereignty, stability and regional security, may overpower other norms on the environmental regime agenda despite rhetoric to the contrary. Justice for oppressed or subordinated collect-ivities will be there of course, but its 'bodily weight' will have been removed by the time it becomes codified into the rules of the game. The rules of the game will term it humane treatment on a fair, efficient, cost-effective basis.

If we were broadly to characterize the formation of most environmental regimes as part embodiments of an institutionalized justice, how would they be characterized? Perhaps as governance systems of minimalist justice, both in form and effect, for subordinated collectivities that deflect attention away from interrelated oppressions and the accumulated multiplying anomalies and contradictions in contemporary non-ecological humanist interventions. The norms and decision-making procedures would typically protect the broad section of international masculinist oligarchies consisting of executive actors from transnational companies, the military, multilateral banks and their major financiers and media conglomerates.

From this view, if we were to advocate the formation/building of an environmental refugee regime, unless we were seriously to problematize its form and modus operandi, we would be ultimately minimizing justice for environmental refugees by staying within the 'rules of the game'.

In order to maximize justice both in form and effect for environmental refugees, it would seem then that a transnational institution or regime is needed for environmental refugees with a capacity to regulate interactions formerly considered 'outside' the realm of regime norms. Rather than focusing myopically on regulating environmental refugees, concerned ecological state and non-state actors, together with displaced people can begin by affirming in a variety of contexts, the human rights and cultural identities of people who become environmental refugees. As for migration regulation policies, perhaps regulating the mobilities and transactions of unscrupulous transnational corporations and their functionaries is a good start.

Perhaps there will come a time when unscrupulous transnationals could have their charters revoked and their assets sold and redistributed to an ecological institution working in the interests of the most exploited classes: '[a]nimals, trees, water, air, grasses' (Snyder, cited in Swanson, 1991) and environmental refugees. Perhaps one day we humans will dwell with humility, in a place I would call this big, big, awesome, planet – our earth.

Notes

1. Within Hardin's lifeboat analogy (see Hardin, 1972) ethical choice collapses into dualisms and questions of asylum. Applied to the environmental refugee situation, the issue becomes one of whether to allow refugees from developing countries to board the lifeboat of developed countries, or

whether to promote sustainable development in their homelands. Debates over issues such as migration rights are ignored.

2. I borrow this metaphor from the Christine Sylvester's feminist art (see Sylvester, 1997: 19).

3. My argument here, contra Haraway, is that wide-ranging generalizations can be situated knowledges.

4. I use this term to designate a critical practice which transgresses academic disciplinary boundaries drawing on a variety of multi-discursive genres to bring an issue to light. Such a practice 'presses the unpleasant questions ... the taboo questions' (Steiner, cited by Henderson, 1995: 3). Feminist border-crossers emphasize important relations which are obscured by androcentric positivist accounts of political terrains and issues. (For examples of feminist border-crossers in the field of 'international' politics see Peterson, 1992; Sylvester, 1994; Pettman, 1996.)

5. There are important differences between and within each discursive practice and some do not sit well with each other, e.g. postmodern feminism with ecological feminism. For a misreading of postmodern feminism from a particular ecofeminist position, see Salleh (1997). For a misreading of ecofeminism from a particular postmodern feminist position see Ferguson (1993).

6. See Sylvester (1994) for a discussion of Maria Lugones's methodological approach and an elaboration of the theme.

7. I am using the term 'you' here as a technique of sensitization, rather than as a process of 'othering'.

8. See Hayter (1989). There are multilateral and bilateral sources of official aid. Multilateral sources include the World Bank as well as regional banks like the Islamic, Inter-American, Caribbean and African banks. On the subject of official aid there are various normative positions. The passage quoted here does not imply a termination of official aid assistance.

6
Environmental Accountability and Transnational Corporations
David Humphreys

Introduction

At the start of the twenty-first century most foreign-based transnational corporations (TNCs) have more power over governments than do the citizens of the countries where they invest. With the growth over the past three decades of corporate power – described by Vidal as the consequence of 'corporate globalism', an ideological challenge to social welfarism (1997a), and by Korten as 'corporate libertarianism', where the rights and liberties of corporations are placed ahead of those of individual citizens (1995: 74) – the task of bringing corporations under democratic control is one of the most pressing challenges for global governance in the twenty-first century.

The ethical case for regulating TNCs is both an environmental and a social one. TNCs have a key role in centre–periphery relations, being both the principal vehicles by which the centre imposes its ecological footprints on the periphery and a mechanism for extracting profits from periphery to centre. TNCs, many of which have an annual turnover greater than the Gross National Product of most developing countries, produce most of the world's atmospheric pollution (leading to global warming, acid rain and ozone depletion), marine pollution (leading to the destruction of marine life) and toxic wastes, and they drive environmentally destructive activities such as mining, overfishing and deforestation.

The issue of regulating corporations has been on the international agenda since the 1970s. The United Nations Centre on Transnational Corporations (UNCTC) was given the task of drafting a code of conduct for corporations. However after nearly two decades of negotiations the

United Nations Conference on Environment and Development (UNCED) failed to agree upon the need for a code, let alone the principles it could contain. In 1993 the UNCTC was closed and its activities downgraded and transferred to the United Nations Conference on Trade and Development's Division on Transnational Corporations and Investment. However, these developments have not, as those who resist calls for corporate regulation would wish, led to a closure of the debate. Corporate accountability remains a salient issue in international civil society.

This chapter provides a rationale for the environmental regulation of TNCs. It is argued that while much of the economic power tradition-ally regulated by states now resides with corporations, arguments that this is in some way inevitable due to globalization are unconvincing. Four contemporary models for corporate regulation are then examined. For different reasons these models have failed to promote the values of social justice and ecological sustainability in corporate practice. The chapter concludes by arguing that equitable and ecologically sustain-able global governance can be realized if the democratic state assumes a more proactive regulatory-legislative role and if states and citizens work to promote the cosmopolitan democracy project (see Chapter 13).

Rationale for the regulation of transnational corporations

Since the 1972 United Nations Conference on the Human Environment in Stockholm there has been a growth in national legislation and a pro-liferation in the number of intergovernmental environmental regimes. However a second process has accompanied this increase of state activity at the national and international levels. This is the shift in the nature of the economic activities of the state following the breakdown of the post-war tripartite Keynesian consensus between the state, business and trade unions and the subsequent promotion of market values, first in the United States and Britain under the Reagan and Thatcher governments, and then in Western Europe. The shift to the private sector has been more widespread than is sometimes recognized, and has occurred to dif-fering degrees in the former communist countries of Europe, China and many developing countries. Central to the neoliberal ethos of what at the time was known as the 'New Right' was the roll-back of the state from the public sector on the assumption that the collective public good was best served by market forces rather than state intervention.

While there is a tension between these two processes, the withdrawal of the state from many areas of economic activity on the one hand,

and increasing environmental regulation and law-making on the other, in fact both processes have resulted in TNCs gaining a growing share of economic and political power.

First, the retreat of the state has created political space at the national and global levels, some of which has been occupied by environmental non-governmental organizations and social movements, but most of which has been colonized by corporate interests. The privatization of public utilities such as gas, electricity and railways has not only empowered the domestic private sector but has also brought many public assets into foreign ownership. A further dimension of the power-shift from the state to TNCs concerns intergovernmental organizations, and has led to what may be termed the 'privatization' of the United Nations. While the term 'privatization' is commonly used to denote the transfer of the means of production from the public (state) sector to the private (market) sector, privatization more broadly defined denotes a process by which corporations are increasingly influencing decisions and activities that are nominally the prerogative of intergovernmental fora. Examples of increased corporate influence within the UN system include telecommunications corporations (International Telecommunications Union), the shipping industry (International Maritime Organization), the timber trade (International Tropical Timber Organization) and the pharmaceutical industry (World Health Organization) (Lee et al., 1997). Such patterns of influence reinforce global governance that promotes the influence of certain economic elites rather than the values of equity and social justice.

Second, the growth of environmental regulation and the negotiation of international regimes has led corporate leaders into greater engagement with politicians. Corporate responses can be placed between two poles. First are those business elites that have developed environmental concerns, but who find it difficult to survive in a global system driven by economic rationalism. Second are those elites that view the environmental lobby as a threat and have responded by trying to peripheralize the environment in the political process. Until the mid-1980s, TNCs tended to reject the need for environmental regulation and to resist implementation of regulations. When it became clear that the environment had become a permanent agenda item, this reactive strategy of responding to political outputs switched to a more proactive strategy whereby TNCs sought to influence the political process from source (Falkner, 1997).

In response to calls for regulation, TNCs have lobbied politicians by forming international associations and federations. In 1990 the

Business Council for Sustainable Development was created to present a corporate perspective to the UNCED. This body worked with the International Chamber of Commerce (ICC) and is credited with derailing the UNCTC's proposed code of conduct (*Third World Resurgence*, 1992: 22). In 1993 the World Industry Council for the Environment was formed by the ICC. In January 1995 the two associations merged to form the World Business Council for Sustainable Development. A further corporate front group is the Global Climate Coalition, a group of energy corporations including Amoco, Chevron, Du Pont, Exxon, Shell, Texaco and Union Carbide which has resisted stronger provisions in the Convention on Climate Change (Rowell, 1996: 86).

Such alliances seek to influence politicians and legislators, usually behind the scenes, to promote free trade and enhance the economic interests of corporations in international law. The European Round Table of Industrialists, which includes Daimler-Benz, Fiat, Shell and Siemens, has lobbied politicians to drive European economic integration to the preferences of European corporations, namely low environmental standards, low inflation, low public expenditure and low wages. European corporations are credited with drafting the original proposal for a Single European Market (Vidal, 1997b: 31). During the GATT Uruguay Round, Unilever, Hoechst and Ciba Geigy targeted the European Union while the United States delegation was influenced by the Intellectual Property Coalition, including Pfizer, Monsanto and Du Pont. Their efforts were noticeable in the drafting of the Agreement on Trade Related Aspects of Intellectual Property Rights (TRIPs) which extended the GATT mandate from purely trade issues to intellectual property rights, including patents for life forms developed by biotechnology (Purdue, 1995). US corporations, including General Motors, General Electric, United Technologies and IBM, campaigned successfully for the negotiation of the North American Free Trade Agreement (Korten, 1995: 145). To Hempel, corporations, now effectively beyond the reach of many national governments, are 'quietly diminishing the power of governments to determine how trade shall be conducted' (1996: 190–2).

Although TNCs may claim that buoyant domestic markets, political stability and a well-developed economic infrastructure are important considerations in where they invest, in the global market place corporations tend to gravitate to countries with low environmental standards. Their power to do so is enhanced by acquisitions and mergers leading to the creation of global monopolies, with entire economic sectors increasingly dominated by a few corporations. Recent examples include the merger of aircraft corporations Boeing and McDonnell

Douglas (June 1997), the merger of Chrysler and Daimler-Benz (May 1998), the takeover of Amoco by British Petroleum (August 1998), and the merger of French drugs company Rhône-Poulenc with German rivals Hoechst (November 1998).

The growing power of corporations has important consequences for world social order. To Lasch, whereas at the start of the twentieth century the greatest threat to social order was the 'revolt of the masses', by the late twentieth century the 'revolt of the elites' was the greater threat. Economic elites know no local or national loyalties, their status and privileges are tied to business, and they have more in common with other members of the transnational elite than with people in their own country (Lasch, 1995: 45). The result is an 'overclass' whose members are removed from local social and environmental concerns and who 'see themselves as world citizens, but without accepting any of the obligations that citizenship in a polity normally implies' (Lasch, 1995: 47).

Below I examine four recent models for regulating corporations and their weaknesses. Then I outline possible mechanisms for the democratic environmental regulation of corporations.

Contemporary models for the regulation of transnational corporations

Four types of model have so far been launched for regulating TNCs: codes drafted by corporations; codes drafted by civil society actors; codes drafted by corporations and civil society actors; and the code of conduct of the United Nations.

Codes drafted by corporations

The majority of corporations favour voluntary guidelines. The best-known statement is the Business Charter for Sustainable Development (BCSD), a declaration of 16 principles produced by the International Chamber of Commerce (ICC, 1991). The charter can be seen as an attempt to move the regulatory debate from command and control instruments towards voluntary codes (Welford, 1996: 66), and an effort to 'reduce the pressures on governments to overlegislate, thereby strengthening the voice of business in public policy debates' (Thomas, 1992: 326). Certainly corporations were more willing to declare support for the BCSD than the United Nations code then under consideration. However there is no independent scrutiny to ensure that corporations implement the principles. As a voluntary framework the BCSD has not provided an ethical basis for corporate conduct.

Codes drafted by civil society actors

At the UNCED Global Forum NGO campaigners produced their own Alternative Treaties. Treaty 16, the 'Treaty on Transnational Corporations: Democratic Regulation of their Conduct' notes that:

> Presently there is no force, governmental, intergovernmental or non-governmental which is capable of monitoring or regulating the activities of these large corporations...recent events show a trend to give more power to TNCs.
>
> (NGO Alternative Treaties, 1992, para. 2)

A weakness of codes drafted by civil society is that corporations have almost entirely ignored them so that their contribution to global environmental governance has been negligible.

Codes drafted by corporations and civil society actors

While ignoring the NGO Alternative Treaties, some corporations have worked on standard-setting in cooperation with non-state actors, including environmental NGOs. In 1989 the Coalition for Environmentally Responsible Economies (CERES) produced ten principles, the Valdez Principles (named after the *Exxon Valdez* disaster), renamed the CERES principles in 1992. CERES is an alliance of corporations, trade unions, religious groups and environmental NGOs. By 1998 over fifty corporations had endorsed the principles, including Sun Company (a petroleum refining company and the first Fortune 500 company to endorse the principles) and General Motors (the largest US-based corporation).

CERES-endorsing companies are committed to reporting annually on their implementation of the principles and to disclosing relevant information to affected parties. While there is an element of accountability to non-corporate interests (Smith, 1995: 116–22) the CERES principle involves an essentially self-regulatory approach. Corporations retain the option of non-endorsement. Both the principles and endorsing corporations have their critics. The principles carry a disclaimer that they 'are not intended to create new legal liabilities, expand existing rights or obligations, waive legal defences, or otherwise affect the legal position of any endorsing company ...' (CERES, 1992).

Lewis argues that one of the reasons why Sun Company endorsed the CERES principles was to gain credibility when lobbying the US Congress against costly environmental legislation (1995: 123–32). Corporate endorsement of a code need not necessarily therefore provide proof of environmental credentials.

The code of conduct of the United Nations

During the UNCED preparatory stages a proposal by the Group of Seventy-Seven Developing Nations and China to establish a code for corporate responsibility based on the draft of the UNCTC (United Nations, 1992) failed to gather sufficient support. Environmental groups report that 'Corporations enjoyed special access to the Secretariat, and the final UNCED documents treat them deferentially' (Bruno, 1992) and that 'discussion within UNCED on the role of business and industry was in effect sidetracked by the creation of the Business Council for Sustainable Development which ... made no suggestions for concrete action plans' (WWF, 1992: 10). The draft UNCTC code is unlikely to be reactivated following the downgrading of the UNCTC in 1993 (see p. 88).

These four models have met with differing degrees of failure. Voluntary codes drafted by business do not contain principles that may benefit the environment if they are also detrimental to the profit and loss accounts of corporations. Codes drafted by environmental groups within civil society have been ignored by the corporate sector. The CERES principles involve an element of external accountability, although they are voluntary and their long-term effectiveness is uncertain. Corporations have successfully resisted a United Nations code based on the UNCTC proposals. The next section will argue that the effective democratic regulation of TNCs requires a stronger role by the state and the democratic reform of international institutions.

Towards the environmental governance of transnational corporations

One view of economic globalization is that it is inevitable: a 'natural' trend that governments can do little if anything to affect. According to this view, the state can no longer exercise control over its economy, and its contribution to economic prosperity is essentially limited to two areas. Nationally its role is to make the economy as competitive as possible through supply-side measures, such as education and training, while internationally its role is to attract corporate investment through, for example, tax breaks. Regulation of corporate activities by the state is not an option in a globalized world.

This view does not stand up to scrutiny. First, many corporations continue to be 'nationally embedded' in one country and are thus in principle subject to regulation by their home state (Hirst and Thompson, 1996: 98). Second, while many corporations have in excess of 50 per cent of sales, employment and assets diversified outside the home state (Bagchi, 1994: 28), all TNC operations are grounded somewhere in a state territory, and are thus subject to national law. Third, under international law, no promises made by a state to a TNC, even those in a state–corporation investment agreement, can exempt the TNC from the law of the land, nor the state from the duty to uphold the public interest (Wiener and Kennair, 1998).

The case is argued here that the state retains the option of regulating most corporate activities. In so doing, a conceptual distinction is drawn between those transnational forces that operate at the sub-state level, and which can be drawn back under public control, and those that operate 'above' the state. We will return to the latter below.

The state has three options with respect to transnational forces operating at the sub-state level. The first is completely to surrender control. The second is to pass control to an international forum. The third option is 'redomestication', the process of 'reigning in' and resubjecting transnational forces to domestic control (Wiener, 1997: 52). One mechanism of redomestication is to require all corporations – national and transnational – to have a public charter in order to operate. In order to empower local communities, charters should be issued not at the national level but, in line with the principle of subsidiarity, at the lowest administrative level practicable. Public charters should be subject to periodic and transparent review by publicly accountable authorities and should be suspended or revoked if a corporation fails to act in the public interest. Mandatory financial penalties should be imposed on firms that delay implementing environmental standards within the stipulated time frame.

The severance of the business–politician relationship is an essential prerequisite for redomestication, indeed for sustainable global governance. Corporations should be denied any input to the legislative-political process and should be prohibited from donating to political parties (which should be funded from public monies). Politicians are first and foremost the servants of the public rather than of sectional interests and should be barred from holding company directorships, thus removing the pressure on them to promote business interests in legislation. While individual corporations should be legally account- able, they should be denied the rights that are granted to individual people as citizens. Korten attributes the growth of corporate power in the US to an 1886 Supreme Court ruling that a corporation is a 'natural person' under the constitution:

> The subsequent claim by corporations that they have the same right as any individual to influence the government in their own interest pits the individual citizen against the vast financial and communi- cations resources of the corporation and mocks the constitutional intent that all citizens have an equal voice...
>
> (Korten, 1995: 59)

National regulation alone is a necessary but not a sufficient condi- tion for constructing an ecologically sustainable global governance. As long as different states have different environmental standards, national legislation can only be successful in regulating TNC activity within a given country (where the activities of a corporation in coun- try A affect the environment of country A). Such legislation will not be successful where corporate activities have transnational effects (where the activities of a corporation in country A affect the environment of country B, usually through a causal chain of intermediary agents, some of them in other countries). Furthermore the individual states enacting the higher environmental standards, and the corporations affected by them, would be placed at a comparative disadvantage economically. The first countries to enact strict regulations would face economic costs such as loss of foreign direct investment. Corporations presented with the higher production costs of environmental regulation would face pressure to relocate to countries with lower environmental standards. As entities driven by economic growth and profit, corporations cannot be expected to adhere to environmental standards if loopholes remain. Even corporations that are not morally disengaged face very real con- straints. Kemcor Australia captures the essence of the problem:

> Kemcor's program to reduce greenhouse gases is a slight one – not because the company does not accept the role of CO_2 in global warming, but because it cannot afford to be economically irrespon- sible by acting alone.
>
> (Suggett and Baker, 1995: 170)

TNCs will always seek to escape high environmental standards unless their business rivals are bound by the same rules – what is sometimes euphemistically referred to as 'a level playing field'.

Hence a harmonization of international standards is necessary to ensure that corporations adhere to the same environmental standards in all countries. The question then becomes, how can this best be achieved? International environmental law can play a role in negotiating global norms which states then internalize at the national level. One option is that when states negotiate an international legal instrument, they also negotiate an annex stipulating how the instrument applies to TNCs. However at present it is far from clear that international law has the capacity to provide a force for environmental conservation. The close relationship between business and political elites in almost every country has led to the construction of the national interest as economic growth and development (Evans, 1998: 208). The need to demarcate and separate the political and business domains again becomes apparent. International law can only become a force for environmental conservation when the ethics of environmental and social justice, rather than the criteria of economic rationalism, provide the basis on which international law is made. Such a project would involve international law addressing 'the political, social and economic context of [environmental] degradation' (Evans, 1998: 223).

Cosmopolitan democracy may provide a second force for the harmonization of international environmental standards. Cosmopolitan democracy also provides a conceptual framework for addressing transnational activities that operate 'above' the state and which cannot be redomesticated at the national level. For example, the degradation of the oceans and the atmosphere (the global commons) is not easily addressed by national action. More fundamentally, the cosmopolitan democracy project seeks to address the structural economic and political forces that constrain life chances and environmental sustainability, particularly in many developing countries through, for example, poverty, high external debts and net South-to-North financial flows. Cosmopolitan democracy thus provides an agenda to realize and to reify the values of socially just and ecologically sustainable global governance.

The theory of 'cosmopolitan democracy' (see Chapters 12 and 13 of this volume) seeks to bring transnational processes under democratic control and is based on the premise that self-determination can no longer be situated entirely within the boundaries of the state (Held, 1997: 260). Cosmopolitan democratic law is defined as 'democratic public law entrenched within and across borders' and is 'a domain of law different in kind from the law of states and the law made between one state and another, that is, international law' (Held, 1995: 227). It calls for new transnational democratic institutions that would coexist with the international state system, but with the power to override states where activities have transnational consequences (Held, 1995: 264). The initial impetus for cosmopolitan democracy could come from liberal democracies and civil societies committed to democratic principles. Short-term objectives of the cosmopolitan model include the establishment of a second chamber to the UN General Assembly (possibly directly elected by the peoples of the world), compulsory jurisdiction before an international court and economic coordinating agencies at the regional and global levels. The longer-term objectives include establishing a global parliament and regional parliaments with the authority to call to account transnational economic actors (Held, 1995: 266–86).

TNCs are, it should be emphasized, just one type of transnational actor that would be affected by cosmopolitan democratic law. How might they be affected? The theory of cosmopolitan democracy is still in its infancy, although the following speculations are consistent with the cosmopolitan democracy project. First, individual corporations should have no input to the shaping of cosmopolitan democratic law, although individuals employed by corporations and committed to democratic principles would have the same rights of participation and voting as any other individual citizen. Second, shareholder meetings should include a public forum open to all citizens who would formally have the right to call the corporation to account for its environmental policies. Third, cosmopolitan law would eventually establish an international legal framework providing the basis for the harmonization of national environmental standards. Fourth, any jurisdiction problems should be resolved by an International Environmental Court (see Chapter 14).

Transnational corporations are, by their very nature, subject to the national law of more than one state, a factor that already leads to jurisdiction conflicts. The International Environmental Court would have the task of resolving conflicts between two or more bodies of national law or between national law, international law and cosmopolitan democratic law, as well as settling liability claims and awarding damages. However the International Environmental Court would only contribute to long-term environmental sustainability if its judgments, and hence cosmopolitan democratic law itself, were grounded in environmental ethics and the values of justice and equity. Fifth, and as argued earlier, public charters should be issued in line with the principle of subsidiarity and subject to periodic and transparent review by public authorities. Finally, cosmopolitan law could provide a legally backed International Convention on Transnational Corporations elaborating principles of environmental sustainability for, and the duties and obligations of, TNCs. TNCs would be required to 'ratify' this convention through their internal mechanisms, such as Boards of Directors, so that it would form part of the company's statutes. The convention would formally codify the obligations that corporations have so far been able to evade.

The Business Charter for Sustainable Development, NGO Alternative Treaties, CERES principles and United Nations draft code provide an indication of some of the principles that an International Convention on Transnational Corporations would contain. These include: the pre-cautionary principle (BCSD, 1991, principle 10; NGO Alternative Treaties, 1992, para. 12); the use of environmentally clean technology (BCSD, 1991, principle 13; NGO Alternative Treaties, 1992, paras. 9 and 10); the elimination of the trade in wastes (NGO Alternative Treaties, 1992, para. 10); the safe disposal of wastes (CERES, 1992, principle 3); the use of Environmental Impact Assessments (BCSD, 1991, principles 5 and 9); freedom of information (BCSD, 1991, principle 15; NGO Alternative Treaties, 1992, para. 8; CERES, 1992, principle 8); and a commitment that corporate employees should not be involved in monetary transactions with public officials (United Nations, 1992, principle 10). In addition there are principles not contained in these four sources which an International Convention on Transnational Corporations would be expected to include, such as full internalization of environmental and social externalities, environmental accounting, independent environmental audits, the polluter pays principle, a recognition of the differentiated responsibilities of corporations, a reversal of the burden of proof with corporations required to prove that new technologies are environmentally safe (rather than that the injured party prove damage) and the use of life-cycle analysis to ensure the safe decommissioning of products and factories.

In order to be effective, with 'effective' defined as the maintenance or improvement of environmental quality rather than the mitigation of environmental decline, any international instrument to regulate corporations will need to be grounded on environmental ethics (and not reflect a median position between corporations and other stake-holders), will need to enforce uniform standards across the globe, and will need to be monitored and verifiable.

Conclusion

Global economic processes have weakened the regulatory capacity of the state, but not to the extent that is sometimes suggested. The principal reasons why unelected and unaccountable corporations have gained economic power are, first, the shift of economic activity from the state to the private sector and, second, the deference of the state to large corporations who to a considerable degree are now able to set the parameters of what constitutes 'acceptable' business practice and 'rational' economic management. The result is a global economy where the interests of ordinary people and the values of ecological sustainability have been rendered subordinate to the pursuit of the corporate interest.

Public accountability involves 'agreeing to follow an ethic developed by representatives of the public' and is 'a norm of responsibility for membership in a community' (Smith, 1995: 120). There is currently no wider dimension to corporate control, which is essentially an internal process differentiated through three layers: boards of directors, senior management and shareholders. However, the state and public authorities not only have both the right and the duty to control their economic and environmental destiny in the interests of their citizens; they also possess the capacity to reclaim the democratic regulation of transnational activities operating at the sub-state level. To argue otherwise is to perpetuate the myth of the enfeebled state. But, as Hobsbawm argues, corporations remain nervous of state power (1998: 6). Hobsbawm refers to the attempts by corporations to promote a Multilateral Agreement on Investment (MAI) through the Organization for Economic Cooperation and Development. The draft MAI prohibits governments from favouring domestic rather than foreign investors and gives TNCs the right to sue governments who erect barriers to trade. In intent and content the draft MAI, which seeks to codify the rights but not the obligations of foreign investors, is the opposite of the International Convention on Transnational Corporations proposed above.

Such a convention is necessary as the transnationalization of economic activities poses a challenge to the democratic regulation of transnational economic forces that cannot be redomesticated. No state acting alone can provide a global ethical framework for conduct, hence the democratic reform of international institutions is necessary. It has been argued that the theory of cosmopolitan democracy provides both an agenda and the conceptual framework for realizing an ecologically sustainable democratic global order. Transnational corporations will have a place in this order, although their sole source of legitimation should be as organizations granted a conditional privilege by publicly accountable authorities who retain the sole right to set ecological and social standards.

Part II
Towards a Global Ethics

7
Towards an Environmentalist Grand Narrative

Arran Gare

Introduction

Modernity is characterized above all by the quest for control: first, the quest for control over nature, but second, control over people to extend control over nature, and third, control over such control. Success in this quest has not only enabled people of European descent to dominate most of the world, but has led most of those societies and civilizations which have not been destroyed by Europeans to embrace this component of their culture. But it is the very success of modernity which has generated a global ecological crisis (Gare, 1996). We now have a global civilization so fixated on how to control things that it has become almost blind to and destructive of the immanent dynamics of the natural and social processes that maintain the conditions for human life. How can we respond to this situation? The response that follows naturally from the culture of modernity is to strive to control more effectively our control over nature and people. As Heidegger observed,

> the instrumental conception of technology conditions every attempt to bring man into the right relation to technology ... We will master it. The will to mastery becomes all the more urgent the more technology threatens to slip from human control.
>
> (Heidegger, 1977: 5)

Now that modern technology is being used throughout the world, we need global mastery of its use. We need global governance.

Can the cause of a problem be its solution? It is this predicament which has provoked scepticism about the whole culture of modernity.

It has led to what Lyotard referred to, defining the postmodern condition, as incredulity towards grand narratives (Lyotard, 1984: xxiv). The grand narratives of modernity have celebrated the present, the 'modern', as a stage on the way from an inferior past to a projected greater future, defining this ongoing transformation as progress. To be contributing to progress is good, and hindering it, bad. All aspects of life, all organizations and institutions, including institutions of government, have been made sense of and evaluated in terms of and then transformed to participate in progress. But progress involves the reduction of everything and everyone, all forms of life and all forms of society, to instruments of this progress. Those who have defended alternative grand narratives to the grand narrative of market-driven economic progress, most notably Marxists, are perceived to be infected by the same domineering tendencies they purport to oppose. All grand narratives are condemned as inherently domineering, homogenizing and oppressive. Eco-feminists and deep ecologists as well as previously despised groups and cultural theorists influenced by poststructuralism, are now suspicious of all totalizing visions of history, and particularly hostile to political action based on such visions.

What is the alternative? To stop trying to control things? To subvert organizations trying to control things? To revert to primitive forms of hunter gathering and agriculture with only local political organization? Or is there some other alternative? My contention is that by invoking the notion of narrative to redefine the problem of modernity, Lyotard's analysis enables us to redefine the very nature of ethics and politics and the relationship between them. This allows us not only to better understand the oppressive features of modernity, but more importantly, to envisage new possibilities for the future. It enables us to create a new grand narrative that could form the basis of an environmentally sustainable civilization with a corresponding global political order.

Narratives, life and action

What is a narrative? Leaving behind its original meaning as a component of rhetoric, 'narrative' now means the mode of discourse that tells a story. First and foremost, narratives are discourses about actions, usually a number of actions, often coming into conflict with one another, with some prevailing over others. Either implicitly or explicitly, actions imply agents with projects and, associated with these, perspectives on the world and their situations within it. So narrative accounts also reveal projects and perspectives; and where more than one actor is involved, narratives often portray not merely conflicts of actions, but conflicts of projects and the perspectives on which these are based. In the case of collective agency, groups can be anything from two people in a particular situation at a particular time, to humanity as a whole over an extended duration. Where collective actions, agents and projects are concerned, the perspectives involved are often very complex and almost always involve dissension.

However, stories are not merely told; they are also lived (Carr, 1991). The present lives of people with their perspectives on the world and ideals and projects can be understood as unfinished stories or narratives, narratives whose plots are yet to be resolved. Paul Ricoeur has noted that there is a complex relation between narratives that are lived and those that are composed and then recounted (Ricoeur, 1984: chap. 1). Life is first of all lived as inchoate narratives that prefigure the stories we tell – we grow up into a world that is already structured and organized by narratives. Secondly, composing stories, creating new 'emplotments', involves configuring life, granting events, actions, agents and perspectives a coherence within the quasi-world these stories project. Reception by people of such stories involves entering these quasi-worlds, distancing them from the narratives they have been socialized into, enabling them to refigure their worlds and their lives. Narratives are lived, critically reflected upon and reformulated through discourse, then reappropriated and lived out by people, both as individuals and as members of groups.

This process is occurring continuously at the level of individual lives, social movements, organizations, institutions, nations, civilizations and humanity as a whole. Stories which people compose on the basis of and as a response to the lives they are living can transform not only individual lives, but the lives of communities, nations and civilizations. Social movements get under way when people begin to construct stories about the disparate efforts to achieve certain goals, and people come to see each other in terms of, and then identify themselves and coordinate their activities through, these stories. So there is a dialectic between life and discourse about life which underlies almost every aspect of human endeavour and every facet of human reality.

As noted, narratives are centrally concerned with action, and as lived, are orientations for action. As Alasdair MacIntyre pointed out, 'I can only answer the question "What am I to do?" if I can answer the prior question "Of what story or stories do I find myself a part?"' (MacIntyre, 1984: 216). To be part of a story is to be part of a larger action, to be participating through one's particular projects in the realization of broader projects. Through constituting and then being sustained by traditions, these broader projects can extend far beyond the lives of individuals. In fact they can extend for thousands of years. Social life involves participating in a complex of stories defining a complex of projects, largely hierarchically ordered with shorter-term projects often being contributions to realizing broader, longer-term projects.

Narratives and politics: truth, justice and the good

While the relationships between narratives and life, discourse and action are very different in different societies, understanding narratives in abstraction from any particular society makes it possible to understand such differences. In particular, it enables us to understand the difference between traditional societies and civilizations, and between earlier and non-European civilizations and the civilizations of Christendom and of modernity.

For the most part, people in traditional societies, without literacy, were never under as much pressure as people in civilizations to legitimate their practices and projects, and their long-term projects were never as fully elaborated or developed. It was in the Ancient World of China, India, Babylon, Egypt, Israel and Greece that such pressure was experienced, and because of the peculiar situation of Ancient Greece, it was here that such pressures were both experienced most intensely and responded to most creatively (Horton, 1967). Through the effort of people forced to confront the relativity of beliefs to societies, abstract notions of justice, truth and the good were first developed.

Generally, people associate the development of these abstract notions with the emergence of philosophy. The introduction of the notion of truth emerged with the distinctions between reality and appearance, and with what is by nature and what is merely customary. This in turn was central to the development of the abstract notion of justice as a standard outside that which is customary and through which any particular practice or action in any society could be judged. The notion of good as it was developed by Plato involved not merely appeal to a standard beyond any particular customs, but the evaluation of all other notions, institutions, practices and actions in terms of this ultimate notion. It involved a functionalist view of the cosmos so that everything, every action and every goal was evaluated in terms of its contribution to the whole. The development of these notions in philosophy corresponded to the lowering of the status of stories. Myths, literature, history and art were all accorded by philosophers (and later scientists) a subordinate status, and this has continued to the present.

However, closer examination of the development of philosophy and then science reveals that narratives or stories were central to this cultural transformation and its further developments, and in fact more fundamental than the development of philosophy. Parmenides identified truth with what is eternal and immutable, and acceptance of this has virtually defined the mainstream of philosophy and later science (with the objects of knowledge being variously conceived as atoms, the forms, the laws of nature, or facts and their logical relations). However, what sustains this definition of philosophy (and subsequently science) is a narratively constituted tradition of inquiry.

In contrast to those who constructed myths, each philosopher from Anaximander onwards defined his own views in opposition to his predecessors, at least implicitly casting earlier philosophers into a historical narrative. This was revealed most clearly in Aristotle's *Metaphysics* in which Aristotle justified the position he defended by casting all past philosophy from Thales to Plato into a historical narrative in terms of which the achievements and limitations of each philosopher were judged from the perspective he was defending. Claims to 'truth' are only sustainable through such narratives, and this is the case not only in philosophy but also in mathematics, logic and science. To become a mathematician, a logician or a scientist involves being socialized into an extended narrative which defines what has been achieved in the past (what is taken as 'true'), what is problematic and what needs to be done in the future.

More significant than the development of the notion of truth in philosophy was the development of a notion of truth in stories themselves. This was associated with the differentiation of stories into histories and fictions. The discourse of history lays claim to truth, and this was the condition of narratives sustaining traditions of inquiry pursuing the truth. However, fiction also lays claim to truth of sorts, and Greek drama was evaluated accordingly. Closely associated with this notion of narrative truth was the development of the notion of narrative justice. The most important concept of justice developed at this time (although opposed by Plato) was 'giving people their due'. While this has been upheld and defended by philosophers from Aristotle to Aquinas, its greatest development was in narratives.

Concern with justice so conceived became central to almost all sub-sequent narratives, whether historical or fictional. Historical narratives are judged not only according to their truth, but according to whether they do justice to all actors (and those acted upon) in the story; that is, whether they acknowledge what people are, what they have endured, what are their potentialities and what they aspire to. Recent attempts to revise history to acknowledge the significance of women, people subjugated by European imperialism, and illiterates and subalterns, exemplify this passionate concern. While such concern is not always evident with fiction, a defining feature of fictional narratives regarded as great literature is that they do greater justice to people than previous narratives.

With this differentiation of narratives, their role in society became much greater and they facilitated much more complex ordering of activities, actions, social movements, organizations, institutions and traditions than had previously been possible. With the synthesis of Greek and Hebraic thought, the Good of Plato was identified with God and reconceived in historical terms as the end and ultimate goal of his-tory. This meant that the functionalism of Plato was reconceived in historical terms, so that activities and institutions have to be seen as contributing to an organic society and an organic cosmos. More than this, all actions and goals, to be justifiable, have to be shown to con-tribute to the realization of the ultimate goal, the reunification of the worthwhile elements of humanity with God. The defence of any par-ticular activity, project or tradition then involved referring it to broader stories and thereby broader projects, and ultimately to the broadest story and project of all. The first grand narrative was born. Following this, there has been a tendency for the narratives by which people define themselves, their traditions, institutions and organizations to be organized as a nested hierarchy coordinated by a dominant narrative defining the ultimate goal of humanity and the cosmos.

The grand narratives of modernity, while rivalling each other, are transformations of this original grand narrative. This is most clearly evident in the alternative modernist grand narratives associated with Hegelian thought. In this tradition, the ultimate Good, whether con-ceived of as the fulfilment of the Absolute or as Communism, is to be realized at the end of history. Historical progress is progress in justice, that is, progress in recognizing people's full potential as rational, cre-ative agents in the political, legal and economic institutions of society. Justification for this view is achieved by casting all past forms of life into a coherent narrative from the ultimate perspective of the Absolute as embodied in the most highly developed society, or from the perspective of a projected communist society in which all the creative powers of humanity have been fully developed and recognized as such. All previous human societies and all non-human life forms have then been judged according to their contribution to realizing this final stage of history.

The dominant grand narrative of modernity differs from this in only minor ways, although these separate it further from the original Christian grand narrative. The ultimate truth pursued by science is the explanation of everything in terms of immutable, mathematically expressible laws, which at the same time will facilitate the complete technological domination of nature, of people and society. The ultimate good towards which all nature and all humanity is developing through an unrestrained struggle for survival, the freeing of markets and the full development of science and technology is the conversion of everything, every person and every organization to efficient cogs of the global economic machine. In this scheme the notion of justice is transmogrified through contractarian notions of rights, utilitarianism and Social Darwinism into a justification of the free market. 'Justice' has become the right of the affluent to exploit nature and other people to the maximum without being burdened by responsibility for the losers in the struggle of all against all – the people who have been rejected as unsuitable components of the economic machine.

The powers of grand narratives

It can be seen from this analysis of the emergence of grand narratives both why they have the power they do, and why postmodernists such as Lyotard are hostile to them. To begin with, their power derives from their capacity to orient people, to provide them with a far stronger identity than was possible before their emergence, to enable far more complex coordination of their activities in organizations and institutions and to help them to work towards goals far beyond the lives of individuals or communities. This means that those societies that embraced and organized themselves through such grand narratives have prevailed over those which did not. Until recently the world was dominated by the struggle between two branches of European culture embodying alternative grand narratives; now there is only one (with minor variations). Globalization in its present form, the denouement of the grand narratives of modernity, amounts to the domination of the entire world by this branch of European culture.

If this is the case, isn't the only possible response to the ecological crisis to reject grand narratives? Shouldn't we undermine them as much as possible and promote a subversive 'rhizome' politics eschewing any overall strategy, direction or coordination (Deleuze and Guattari, 1988: chap. 1)? Postmodernists have been sufficiently influential now to evaluate such tactics. They have been disastrous. The dominant grand narrative of economic progress through liberating the market, fostering a struggle of all against all and reducing all institutions, including universities, to means to control nature and people and to legitimate this struggle of all against all, has been left virtually uncontested to dominate as never before. Opponents of this, deprived of any sense that their own efforts are contributing to a longer-term goal, without a larger story to dignify and give significance to their lives, have become almost totally ineffectual. Not totally ineffectual, however. They have helped undermine or tame all the power centres, partially autonomous cultural fields and reformist political parties that had provided some countervailing force to the logic of the market.

Grand narratives and cosmologies

If postmodern politics has proved ineffectual then it is necessary to again consider alternative grand narratives. The grand narratives that have come to dominate European civilization and thereby the world were intimately associated with particular perspectives on the world, with perspectives based on a Platonist cosmology. While modernist grand narratives have transcended the Neo-Platonism of early Christianity and feudal society, both the dominant narrative based on scientific materialism and the alternative developed by Hegel and Marx, still have their roots in Neo-Platonism (Gare, 1996: chaps 4, 5, 9). It is as a consequence of this that they tend to view the whole of history from what is taken as the one, absolute and final perspective – that which reveals reality itself – and to evaluate everything in terms of its contribution to achieving some final end.

An alternative philosophy, a philosophy which has always had a prominent oppositional role in the human sciences and which is now becoming increasingly influential in the natural sciences, is provided by Heraclitus (Gare: 1996, chaps 13–15). For Heraclitus reality is in flux; each identifiable individual is only a relatively stable entity maintaining itself within this flux and there are no absolutely immutable beings that can be known with absolute certainty. This means that the notion of truth developed by Parmenides and appropriated by Plato is inappropriate to describe the goal of inquiry. More appropriate to an infinitely complex and changing reality consisting of self-ordering patterns of activity would be the notion of understanding. Understanding can be endlessly deepened without ever reaching final point, and different perspectives can advance understanding in different ways (Gare, 1996: chap. 12). Correlatively, the notion of an ultimate good, the totality in relation to which all else must be judged, has no place in such a cosmology.

The notion of understanding without recourse to some notion of an Absolute in relation to which all particulars must be evaluated allows for the full development of the notion of justice as giving not only people, but everything else, their due. Understanding immediately involves evaluation and appreciation of the intrinsic significance of what is understood. It is possible to appreciate that individuals, whether human or non-human, have an intrinsic significance quite apart from their contribution to anything else, and to evaluate totalities in terms of whether and how they define particular individuals and allow them to flourish. Justice involves fully appreciating both the intrinsic significance and significance to each other of every individual, present, past and future, taking them into account in the way we live, the decisions we make and the actions we engage in. The effort to do this is itself an unending quest with no finality.

What status would narratives be accorded in a Heraclitean world? Far greater than in any society dominated by Neo-Platonic thinking. In fact the relative status of narratives on the one hand and mathematics and logic on the other for understanding the world would have to be reversed. If the world is in flux, if it is a creative process of becoming, then mathematics and logic are ultimately incapable of grasping reality. While recognizing their achievements, they must be held to be able only to grasp derivative aspects of reality, those relatively stable islands of stability within the flux where there is no genuine creativity involved. Narratives, on the other hand, presuppose a world of creative becoming with a diversity of often conflicting agents, actions and processes in which the future is uncertain until it is realized. And narratives evaluate these individuals as they characterize them. Contra the grand narratives which have dominated European civilization and which now dominate the world, narratives by their nature do not presuppose there is some definite and inevitable end to any story and do not evaluate individuals only in terms of their contribution to realizing some particular end.

Polyphonic grand narratives

Not only do narratives not presuppose a definite, inevitable end, they do not presuppose that they must be told or understood from one absolute perspective. Mikhail Bakhtin has noted that as well as mono-logic narratives there are polyphonic narratives: that is, narratives which give a place to a plurality of voices without presupposing that any of these has the one, true perspective (Bakhtin, 1984). Polyphonic narratives allow diverse perspectives to be brought into relation to each other and to contest each other, and allow that there are other perspectives not yet represented or considered. Such narratives are brought down to the level of those receiving them. Receivers of narratives can then relate their own perspectives to those represented, to appreciate that their own perspectives are being contested, but also that they can contest the represented perspectives. Monologic narratives, which assume one, true view of reality being conveyed to the reader, or through which people's lives can be understood, are not in any sense 'normal', but can only exist by suppressing the possibility of different voices.

It is through the suppression of other voices and other perspectives that other people and the world generally are reduced to mere instruments. As Bakhtin noted:

> With a monologic approach (in its extreme or pure form) *another person* remains wholly and merely an *object* of consciousness, and not another consciousness .... Monologue manages without the other, and therefore to some degree materializes all reality.
>
> (Bakhtin, 1984: 292f)

Polyphonic narratives involve both representing and respecting other consciousnesses as beings with equal rights and equal responsibilities.

So it is not grand narratives as such which reduce everything to instruments to be evaluated in terms of their contribution to the projects of these grand narratives, but monologic grand narratives with their roots in Neo-Platonism. Polyphonic narratives, construing the world as active, as a world of actions and processes with their own dynamics and ends and allowing diversity of perspectives, provide the basis for the full development of the quest for narrative truth and justice.

Could such polyphonic narratives enable people to orient themselves for concerted action? Or would they lead to the same kind of relativism fostered by postmodernists? Polyphonic narratives are not only not relativistic. They provide the means to overcome relativism. It is through such polyphonic narratives that the truth and justice of diverse perspectives are brought into relationship, questioned and judged. Despite the tendency of philosophers and scientists to claim that finally they have discovered the ultimate truth, the justification of beliefs involves producing narratives from the perspective of the belief, showing how through it the achievements and limitations of alternative beliefs can be understood (MacIntyre, 1977). Such narratives are not monologic but polyphonic, and far from closing off further enquiry, such narratives cannot avoid placing in question any belief or perspective being defended. That is, while justifying the provisional commitment to such a belief or perspective as against previous beliefs, they open the possibility and invite further enquiry to transcend this belief in turn. What is called for is the development of diverse perspectives to reveal as many aspects of the world as possible and to reveal which of these perspectives is most adequate. And adequacy pertains not only to truth, but to issues of justice. Polyphonic narratives are pre-eminently concerned with the justice of the different perspectives and actions of the people and other beings they construe or represent.

Towards an environmentalist, polyphonic grand narrative

With this analysis of narratives and their relation to life we can now return to the issue with which we started: the modernist grand narrative of mastery over the world and the incredulity this has engendered towards all grand narratives.

It is clear that there is every reason to be incredulous towards modernist grand narratives. These narratives underlie the drive to reduce everything and everyone to cogs of the economy. And it is clear that by failing to do justice to the immanent dynamics of individuals, societies and natural processes and to the diverse perspectives on the world of various 'pre-modern' communities, the inevitable by-product has been the global ecological crisis. But there is nothing to be gained by leaving the cultural field to existing grand narratives. If we are to confront the global ecological crisis we need to create a new, polyphonic grand narrative which can do justice to the immanent dynamics and intrinsic significance of the world and all its participants. This narrative must also define the problems confronting us, convince people to embrace them and then orient them for concerted action over centuries to transform this destructive civilization.

This will involve redefining the past, the present and the future. By revealing the present as the product of the destructive trajectory of modernity, we can see ourselves in a crisis in the original sense of the word, a crucial point at which the disease is either overcome, or it continues to its fatal conclusion (J. O'Connor, 1987). Overcoming this crisis will require a new vision or visions of the future. We need utopias which put into question what presently exists and provide goals to aim at (Ricoeur, 1986).

A major task for environmentalists is to imagine a global civilization that could replace that which presently exists, a civilization which sustains the environmental conditions of its existence and allows the world eco-system of which it is part to flourish. Efforts to envisage such a future should be accompanied by efforts to imagine what paths could be taken from the present circumstances of different people, organizations and institutions to reach this future, identifying the crucial decisions which need to be made and crucial battles which must be fought. In this way this image of the future could be integrated with historical narratives, allowing individuals, peoples, organizations and institutions to situate and identify themselves as historical agents, living out an as yet unfinished story projecting this vision of the future. The projected vision should then be continually questioned and revised in the light of changing circumstances, fresh achievements and advances in understanding.

In doing so, an approach is required which takes into account the diversity of perspectives and diversity of voices, and which acknowledges the need to respect these even when opposing them. That is, environmentalists need to struggle for justice for all nature and all people, even those who dismiss or oppose the environmentalists, and who have no notion of justice. Environmentalists should not merely tolerate diversity among their ranks, but encourage such diversity, and not only should they tolerate diversity of cultures throughout the world, but encourage such diversity, while simultaneously struggling to demonstrate in thought and practice the superiority of their own points of view. Only in this way can their points of view be tested and developed. Correspondingly, establishing a global environmentally sustainable political order should be understood not as political mastery over technology, but as the reconfiguration of civilization by a polyphonic grand narrative. This would be associated with a decentralized and balanced system of power fostering a diversity of forms of life and cultures. Only by creating such a plurality of power centres able to constrain each other will niches be provided for people to criticize the existing order and to develop new, more just ways of thinking and living. And only by providing and sustaining such niches will more just ways of thinking and living be able to flourish and eventually prevail.

8
Human Rights and the Environment: Redefining Fundamental Principles?

Klaus Bosselmann

Introduction

From a legal point of view, global governance can be seen as a search for human and environmental rights. While economic globalization necessitates new forms of international control, global governance strives for more. In a world of diluted state sovereignty and market anarchy the search is for human values on which any form of governance could be built. Are there common human values? How much commonality is necessary to nurture the idea of a just global order?

A starting-point for the search is the very experience of dissolving national autonomy. Economic globalization profoundly affects not just local economies, but individual lives of people: it touches upon our sense of security. National economies or governments allocating human rights and environmental resources no longer provide for security. Where distant bodies of no public accountability make such allocations, any sense of security gets lost. We are left with looking (yet again) for the essentials of civil society.

Not surprisingly, the general loss of security goes hand in hand with increased demands for human and environmental protection. Never before have so many people made so many legal claims for both human rights and the environment. Clearly, human rights and the environment are closely related.

But how are they related? Is the environment a mere good or value to be added to the list of individual demands? This is the approach of individual environmental rights. Or is the environment rather a condition of *all* life on earth? This is the ecological approach to environmental rights. The former is consistent with the 'human rights' tradition which seeks to expand the freedom of the human individual and recognizes limitations on that freedom only for the protection of similar freedom of other humans. The latter necessarily requires a limitation of individual human freedom in order to protect the integrity, in some sense, of Nature of which humans are part (as in the *deus sive natura* of Spinoza, in Mathews, 1991). Thus in the ecological approach there is a potential break, even conflict, with the human rights tradition.

This chapter discusses both approaches as they have emerged in international and national law. A human right to a healthy environment has been promoted for more than twenty years and can today be seen as a right *in statu nascendi*. The notion of ecological limitations to human rights, on the other hand, is more recent, not implemented anywhere and in need of some definition (see Bosselmann, 1998: 19–30, 65–79). It refers to the fact that individual freedom is determined not only by a social context – the social dimension of human rights – but also by an ecological context. In ethical terms, the anthropocentric, utilitarian understanding of human rights would be complemented or replaced by an ecocentric understanding which holds that the natural environment has intrinsic value, not just instrumental value.

Linking human rights to ecological ethics could well be the key to a cross-cultural commonality of human values. Beyond the existing multiplicity of cultures and values, there seems to be a deeply embedded sense that life as a whole is to be respected: people ought to respect each other as well as the natural environment.

Considering the far-reaching consequences of such a sense for the vision of a just global order, a debate on ecological human rights is timely. The chapter is organized in three main parts. The first looks at international human rights law and recent developments in individual countries. The second discusses the increasing criticism on anthropocentric limitations associated with present environmental rights, and the final section presents the new ecological approach to human rights.

Current trends in the development of environmental rights

International human rights

Various ambiguities of international human rights place obstacles in the path of any attempt to establish a global order of justice. Following

Weston (1986), some commonly accepted features can, however, be detected:

- human rights represent individual and group demands for the shaping and sharing of power, wealth, enlightenment and other important values in community process. They limit state power;
- human rights refer to a wide continuum of value claims ranging from the most justifiable to the most aspirational. They represent both the 'is' and the 'ought';
- a human right is general or universal in character, equally possessed by all human beings everywhere;
- most are limited by the rights of any particular group or individual and are restricted as much as is necessary to secure the comparable rights of others and of common interest;
- human rights are commonly assumed to refer to 'fundamental' as distinct from 'non-essential' claims.

For an ecological approach to human rights, one of the most important postulates identified above is the limitation or qualification of a human right. International human rights documents employ several different techniques to define the boundaries of rights.

It is also important to realize possible conditions under which limitations of human rights are justified. Typically, three independent components of a justified limitation are used: (1) the limitation must be provided for by law; (2) that law must be *necessary* as opposed to useful or desirable; and (3) it must protect one or more of a limited set of public interests such as national security, public safety, public order, public health, public morals, and the rights and freedoms of others (Sieghart, 1985). Ultimately, the moral or legal weight of a common interest determines the extent to which an individual right may be limited.

Limitations of human rights in legally prescribed circumstances are an accepted practice in international human rights theory. However, consistent with concern for human social ethics and its disregard for the environment, these restrictions reflect only social ethics. Environmental ethics only recently began to influence human rights theory (Taylor, 1998). So, are limitations to international human rights justifiable on the grounds of environmental concerns?

Since 1968 numerous international declarations and statements have recognized the fundamental connection between environmental protection and respect for human rights. In 1968 the UN General Assembly passed a resolution identifying the relationship between the quality of the human environment and the enjoyment of basic rights. The

United Nations Commission on Human Rights adopted a resolution in 1990, entitled 'Human Rights and the Environment', which reaffirmed the relationship between preservation of the environment and the promotion of human rights.

Interestingly, the Ukranian delegation made a proposal for ecological human rights to the Commission on Human Rights. The proposal was innovative, albeit limited to human interests. The Ukranian proposal included: (1) the right to ecologically clean foodstuffs; (2) the right to ecologically harmless consumer goods; (3) the right to engage in productive activities in ecologically harmless conditions; (4) the right to live in ecologically clean natural surroundings; (5) the right to obtain and disseminate reliable information on the quality of foodstuffs, consumer goods, working conditions, and the state of the environment.

In the build-up period to the 1992 Rio Conference on Environment and Development there were many proposals to institutionalize a right to a decent environment. The Draft Earth Charter (1991) is one of them (Hohmann, 1992). However, the eventually adopted Rio Declaration relates the rights issue to the broader issue of sustainable development.

The Earth Charter movement rejected the Declaration's anthropocentric, one-sided approach and advocated sustainability based on the ecocentric principle 'Respect Earth and all life'. Together with a number of international law bodies, including the Commission on Environmental Law, the Earth Charter Commission aims for the endorsement of the Earth Charter by the United Nations General Assembly by 2002 and, at the same time, for the adoption of a UN Covenant on Environment and Development.

Recognition of environmental rights is not limited to environmental agreements. The Organisation for Economic Co-operation and Development (OECD) stated that fundamental human rights should include a right to a 'decent' environment (OECD, 1984: 122). The Charter on Environmental Rights and Obligations drafted by the United Nations Economic Commission for Europe (UNECE), affirms the right of everyone to an environment adequate for general health and well-being and the responsibility to protect and conserve the environment for present and future generations.

A common criticism of environmental rights is that present formulations are too vague and general in terms of their content, scope and enforceability. Often they are perceived as largely aspirational, expressing national goals and intents, rather than justifiable rights (Schwartz,

1993). However, it is also accepted that environmental rights can be 'derived from other existing treaty rights, such as life, health, or property' (Birnie and Boyle, 1992: 192). At present these environmental rights may serve as a 'surrogate protection' against environmental harm (Schwartz, 1993: 375).

Constitutional developments

An important indicator of the development of an environmental human right is the extent to which it has emerged in national constitutions. There are a number of studies which list these constitutional provisions (e.g. Brown Weiss, 1989; Bothe, 1999). They reveal that nearly 60 national constitutions, from a variety of legal traditions, include an environmental human right. Virtually every constitution revised or adopted since 1970 has addressed environmental issues (Kiss and Shelton, 1991). Recent examples can be found in the constitutions of Slovakia, Slovenia, Hungary, Poland and South Africa.

The link between rights and obligations is another recent trend. Most member states of the European Union have recently adopted constitutional changes to include environmental duties and state obligations. They include the Netherlands, Sweden, Finland, Germany, France, Belgium, Luxembourg and Greece.

Michael Bothe (1999: 12) summarizes the present situation in the EU Member States as follows:

> There is a clear trend towards a constitutional recognition of environmental values. On the other hand, this recognition does not necessarily mean that affirmative rights to the protection of the environment are granted. The constitutional recognition of environmental values is a basis for protection against infringements and repressions.

The situation in Germany warrants a closer look. The 1993 amendment to the *Grundgesetz* (constitution) introduced a new Article 20a which defines care for conditions of life and for future generations as a state obligation. This amendment was achieved as a political compromise after nearly ten years of constitutional debate. At the heart of the debate stood the conflict between an anthropocentric and a non-anthropocentric approach. The Greens Federal Programme of 1980 questioned the anthropocentric paradigm by stating that 'humans and their environment are part of nature'. In 1987 the Social Democratic Party proposed a constitutional state obligation ('The natural conditions of life are under special protection of the state') which should avoid anthropocentric reductions. The Greens went further:

> The natural environment, as life condition of humans as well as for its own sake, are under special protection of the state. In conflicts between ecological burdens and economic needs ecological concerns have priority ...
>
> (in Bosselmann, 1992: 197)

In a 'Common Declaration' the German Protestant and Catholic churches stated:

> Any formulation of a state obligation which does not recognize the intrinsic value of non-human life would ... jeopardize the created world as a whole in its biological diversity ... Every environmentally relevant decision has to be based on weighing-up use interests of humans and the intrinsic value of affected non-human life.
>
> (Gemeinsame Erklärung der evangelischen
> und katholischen Kirche, 1989: 37)

Following German unification, a broad alliance of university professors, lawyers and social scientists presented the draft of a new constitution based on the principles of democracy, solidarity and ecology. The new principle of ecology should define both state obligations and individual human rights. The draft constitution postulated that individual human rights are not only defined by 'social limitations', but also by 'ecological limitations'. In 1991, the *Bundesrat* (Upper House) 10 *Länder* (federal states) supported a constitutional initiative by the *Land* of Bremen which called for a state obligation to protect the natural environment 'for its own sake against the effects of free development of human beings' and postulated a general 'ecological limitation' to the individual fundamental right of free development (in Bosselmann, 1992: 201).

The conservative (CDU/FDP) coalition government rejected an eco-centric revision of the constitution. At the end, the Joint Constitutional Commission concluded that it 'was and remains controversial whether the protection of the natural conditions of life should be formulated anthropocentrically ... or not' (v.d. Pfordten, 1996: 288). The constitutional debate is far from over and continues, in particular, in the legal literature (e.g. Steinberg, 1998).

Switzerland and Austria are in the middle of similar debates. In both countries various initiatives call for a constitutional reform based on ecocentrism. In 1992, a new Article 24 of the Swiss constitution established the 'dignity of creatures' (*Würde der Kreatur*) which is to be respected in legislation on genetic engineering. The *spiritus rector* of the constitutional movement, Peter Saladin (1995), argues for an entirely newly designed constitution based on three ethical principles:

- the principle of solidarity (intragenerational justice);
- the principle of human respect for the non-human environment (interspecies justice); and
- the principle of responsibility for future generations (intergenerational justice).

Saladin's analysis of the 'functions of a modern democratic state in an increasingly supranational world' concludes that a solution for collective problems can only be found if the state is revitalized by this 'ethical impulse'.

A constitutional initiative by the Austrian Ministry for the Environment aims for a 'reform of the federal constitution based on ecological principles' (Pernthaler, 1992: 1). This constitutional principle postulates, inter alia, legal rights of nature and future generations and ecological '*Grundpflichten*' (fundamental duties) restricting individual freedom and human rights.

Ecological critique in legal literature

The constitutional debate in German-speaking countries indicates a growing frustration with the anthropocentric tradition of human rights. There is particular concern over the inherent anthropocentricity of an environmental human right. In the view of many commentators the very existence of environmental human rights reinforces the idea that the environment and natural resources exist only for human benefit and have no intrinsic worth. Furthermore, they result in creating a hierarchy, according to which humanity is given a position of superiority and importance above and separate from other members of the natural community.

The human-centred character of an environmental human right leads to a philosophical tension between deep and shallow ecologists. Consequently, some commentators wholly reject human rights proposals (e.g. Gibson, 1990), while others offer a compromise position (e.g. Nickel, 1993).

There are a number of concerns with anthropocentric approaches to environmental protection. They tend to perpetuate the values and attitudes that are at the root of environmental degradation. Further, they deprive the environment of direct, independent protection. Human rights to life, health and standards of living are all determining factors for the aims of environmental protection. Thus the environment is only protected as a consequence of protecting human well-being. An environmental right thus subjugates all other needs, interests and values of nature, to those of humanity. Finally, humans are the beneficiaries of any relief for infringement of the right. There is no guarantee of its utilization for the benefit of the environment. Nor is there any recognition of nature as the victim of degradation.

On the other hand, it can be argued that at least a degree of anthropocentrism is an inescapable part of environmental protection. Humanity may not be at the centre of the biosphere, but as we seem the only species recognizing and respecting morality, we may be able to also include the environment in our existing morality. Such 'enlightened' anthropocentrism would acknowledge that interests and duties of humanity are inseparable from environmental protection. Dinah Shelton (1991: 110), for example, argues:

> [H]umans are not separable members of the universe. Rather, humans are interlinked and interdependent participants with duties to protect and conserve all elements of nature, whether or not they have known benefits or current economic utility. This anthropocentric purpose should be distinguished from utilitarianism.

As a consequence, an environmental human right could be complementary to a wider protection of the biosphere that recognizes the intrinsic values of nature, independent of human needs. As Birnie and Boyle point out, this approach would work to the extent that it successfully 'de-emphasizes the uniqueness of man's right to the environment and conforms more closely to the characterization of this relationship as a fiduciary one not devoted solely to the attainment of immediate human needs' (Birnie and Boyle: 194). Birnie and Boyle see the implications of the issue as being largely structural, requiring the integration of human rights claims within a broader decision-making framework capable of taking into account intrinsic values, the needs of future generations as well as the competing interests of states.

Various compromise positions are offered in the literature. They might be useful in assisting the transformation from an essentially anthropocentric perspective to an ecocentric perspective. However, in the long term the existence of an environmental human right could be seen as self-contradictory. In the age of ecologism the rights discourse is bound to change direction. In future, ecological limitations, together with corollary obligations should be of similar self-evidence as social limitations to human rights.

Attempts to overcome the anthropocentric paradigm in law are plentiful. Among these, the concept of nature's rights has been well documented since its rise to prominence in 1972, following the publication of Christopher Stone's article 'Should Trees Have Standing?' (Stone, 1972). Over the past 25 years the concept has been widely debated among lawyers, philosophers, theologians and sociologists. This debate has led to a wide variety of ethical and legal discourses including: legally enforceable rights for nature; so-called 'biotic rights' (being moral imperatives which are not legally enforceable); moral 'responsibilities'; and 'rightness' (a norm which prescribes a need for a proper healthy relationship between humanity and nature). What is common to them all is an attempt to give concrete and meaningful recognition to the intrinsic value of nature. They differ in how this should be achieved. Some commentators advocate that it should be done within the context of legally enforceable rights, others argue for recognition of values or a status, which requires humanity to take into account the interests of nature.

'Rights talk' is not very popular among non-legal ecologists. Deep ecologists and ecofeminists tend to perceive rights as absolute, static, individualistic and deeply embedded in the anthropocentric (male) paradigm. Lawyers, too, have argued that the concept of nature's rights is tantamount to a 'quick legal fix', which, like many other legal solutions, precludes the deep questions necessary for genuine world change. In the tradition of Marxist legal theory 'rights' can be rejected as an appropriate method of social reform. Likewise, rights may not lead us to change our attitudes to nature. Giagnocavo and Goldstein (1990) reject rights of nature as a 'false claim'. In their opinion, legal 'rights' give the holder some advantages (see Stone, 1972), but this only amounts to valuing by legal institutions, not society at large.

Stone himself recognizes the limitations of his 'rights' theory and in the final pages of his article discusses the importance of a changed environmental consciousness. He states that legal reform, together with attendant social reform will be insufficient without 'a radical shift in our feelings about "our" place in the rest of Nature' (Stone,

1972: 495). Stone has never considered 'rights' as an end in them-selves, but rather as vehicle to transport the ethical debate into the legal discourse. There is, after all, a dialectic between the legal and the political/ethical discourse.

Whether or not law ever implements the concept of nature's rights, the very existence of the debate contributes to the development of eco-logical rights. It has helped to develop consciousness beyond the prevalent anthropocentric ethic by suggesting what to many might have formerly been the 'unthinkable'. Gradual acceptance of moral responsibilities towards nature may lead to a point where we begin to accept the idea of ecological limitations on the exercise of our rights or, more directly, agree to redefinition of the content of certain rights (e.g. property rights).

The ecological approach to human rights

Some environmental lawyers have argued that we should not view environmental issues through a human rights focus, entailing a form of 'species chauvinism'. We should instead think either of nature's rights or of limitations to human rights with respect to the 'intrinsic values' of the environment.

The former idea of rights for nature has been described as the 'strong rights-based approach', the latter idea of intrinsic values as the 'weak rights-based approach' (Redgwell, 1996), which is what is advocated here. There is little reason to believe that an ecocentric turn-around can be achieved just by adding rights of natural objects to the catalogue of human rights. As seen above, there are a number of difficulties with 'rights' thinking, the most important being that we may only foster the very paradigm we are trying to overcome.

The project of ecological human rights attempts to reconcile the philosophical foundations of human rights with ecological princi-ples. The approach is to link the intrinsic values of the humans with the intrinsic values of other species and the environment. As a result, human rights (such as human dignity, liberty, property, and develop-ment) respond to the fact that the individual exists not only in a social environment, but also in a natural environment. Just as much as the individual has to respect the intrinsic value of fellow human beings, the individual also has to respect the intrinsic value of other fellow beings (animals, plants, ecosystems).

The reference to 'respect' for others as the determining factor for individual freedom is not incidental. The literature on environmental ethics and the literature on human rights have certain common ground. Ethical considerations on our relationship with the environment often use the category of respect like, for instance, Paul Taylor in his influential work *Respect for Nature* (1986) or Tom Regan (1992) in his discussion of moral and legal obligations. Indications are that the contemporary ethical debate recognizes intrinsic values as the basis for moral considerability and respect as the basis for personal obligations.

Similarly, we find in human rights theory the concept of 'respect' expressed as the ethical basis for human rights. McDougal, Lasswell and Chen in their standard text on human rights (1980), for example, define respect as the 'reciprocal honoring of freedom of choice'. They suggest that using this universal principle, it is possible to cover all aspects of life requiring protection by formulations of rights. John Rawls's (1971) *Theory of Justice* may not be far from this with its emphasis on a universal principle that needs to be accepted by all in order to create a just society.

The respect for the intrinsic value of life could guide both, the relationship between the individual and society on the one hand, and the relationship between humans and the environment on the other.

Structurally, human rights can be limited by ecological considerations in the same way that they are presently limited, namely by social and democratic considerations. Human rights are not absolute, but subject to a variety of limiting factors. There are general and specific limitations to individual rights. A general reference often used in legislation defining limitations to human rights is the 'reasonable limits prescribed by law as can be demonstrably justified in a free and democratic society'. This phrase is used, for instance, in the European Convention for the Protection of Human Rights and Fundamental Freedoms, in the Canadian Charter of Rights and Freedoms and in the New Zealand Bill of Rights. Typically, any limitation to an individual right has to pass proportionality tests of necessity, lowest possible impairment and balance of conflicting rights.

There is, of course, a considerable variation in how the balance is actually achieved. For example, civil law countries and the United States follow an 'absolutist approach' emphasizing the supremacy of law as defined in the Constitution, and attempting to avoid substantive issues. On the other hand, countries like Britain, Australia and New Zealand follow a 'balancing of interests' approach which attempts to weigh up the various interests. The bottom line, however, is the same throughout all these jurisdictions. It is always concern for the rights of the other members of society that ultimately determines to what extent an individual right may be limited. This bottom line can be referred to as the 'social dimension of human rights'.

On this basis, a closer inspection is possible as to what the essence of human rights and fundamental freedoms is. The essence appears to be the attempt to define the freedom of the individual in interaction with other individuals. Thus, it is the social sphere of human existence which human rights are concerned with, not the biosphere. The biosphere (environment) is presently taken for granted and has no legal quality (Bosselmann, 1995). Historically and systematically, human rights are created to protect citizens against the state, in other words to protect humans from each other; they contain no provision to stop human beings from exploiting non-human beings and fundamentally changing the conditions of life. As long as human rights are not impinged on we are free to destroy the environment and all life around us.

The only existing restriction in this respect is our anthropocentric morality, which may require us not to torture animals, not to turn a beautiful landscape into a moonscape, and to limit genetic engineering to those areas beneficial to us humans. The limits are always drawn by our concern for human welfare to the exclusion of the welfare of other life forms. The dilemma is, of course, that we cannot survive without concern for the welfare of life as a whole. This is the harsh reality discovered by ecology.

To rectify the present situation of grave imbalance there are two options. Either we manage the ethical paradigm shift in society and don't worry about human rights doctrines (we may simply assume that these doctrines would follow sooner or later). Or we promote the ethical paradigm shift at all social levels including the constitutional and legal level.

Without discussing here to what extent the law can make a difference to social behaviour, both the classic views appear to be wrong. The traditional liberal view which holds a profound separation between legal norms and social reality is erroneous. Nor is the Marxist view which denies any distinction between legal norms and social reality appropriate. The law both purely reflects and actively influences the way in which society operates. That is why it matters whether ecological reflections exist in legal norms or not.

For a concept as revolutionary as a non-anthropocentric concept of human rights, the burden of proof is, of course, on those advocating it. What would be the advantage of ecological human rights? Would they make any difference to the real outcome of decisions? One example should illustrate this. It will demonstrate why it would not be sufficient to rely purely on the social dimension of human rights. The example concerns the law related to biotechnology.

At international level, biotechnology became subject to international law through the 1992 Convention on Biological Diversity. Along with a general trend in recent international environmental law, the Biodiversity Convention takes the approach of ecosystem protection (i.e. protecting entire habitats rather than individual species as such). It does so by introducing (in its Preamble) an 'intrinsic value of biological diversity', in addition to 'the ecological, genetic, social, economic, scientific, educational, cultural, recreational and aesthetic values of biological diversity and its components'. This is the recognition of both the (ecocentric) intrinsic values and the (anthropocentric) instrumental values of the environment.

In fact, environmental agreements have increasingly adopted an ecocentric perspective by focusing on the intrinsic value of all life and respect for nature. Examples include the 1991 Protocol on Environmental Protection amending the 1959 Antarctic Treaty, the 1991 Caring for the Earth Strategy (IUCN, UNEP, WWF), the 1991 Hague Recommendation on International Environmental Law and the 32 so-called 'Alternative Treaties' which several hundred non-governmental organizations negotiated at the 1992 Global Forum in Rio. The Global Forum triggered a number of further initiatives resulting in ecocentrically designed international documents. The 1995 Draft International Covenant on Environment and Development, for example, affirms (in its Preamble) 'the essential duty of all to respect and preserve the environment'. Article 2 states: 'Nature as a whole warrants respect; every form of life is unique and is to be safeguarded independent of its value to humanity.'

Article 19 of the Biodiversity Convention calls for the Contracting States to take legislative measures towards controlling biotechnological research activities. The problem is that the Convention, like most treaties, leaves the means of implementation totally to the discretion of states.

At municipal level, several countries have introduced such controlling legislation, among them Germany with its Gentechnikgesetz (Gene Technology Law) of 1990. Such legislation regulates details of the notification and licensing of genetically modified products (e.g. the release of those products into the environment), but it always does so on the basis that there is a fundamental right to conduct genetic engineering in the first place. The principle of free production and sale is the rule; any restrictions are the exception. The burden of proof, therefore, is not on the producer introducing a new risk potential, but on the general public (represented for example by expert commissions such as the Environmental Risk Management Authority in New Zealand or various commissions in the United Kingdom). Whether or not activities of genetic engineering are acceptable, is determined by weighing up social costs and benefits. The problem is that such social costs and benefits are exclusively determined by values of human utility. There are no intrinsic values of ecosystems and their components to be considered.

Quite obviously, there is a gap between the ecocentric approach of the Biodiversity Convention and its implementation through the anthropocentric approach of municipal legislation. To close this gap, one could imagine a simple legislative act to impose the burden of proof onto the producer (or importer) with the consequence that any remaining doubts go against the applicant. However, such radical interpretation of the polluter pays and precautionary principle has not been made anywhere and is unlikely, indeed impossible, to be made on the basis of our current anthropocentric concept of human rights.

Research, development and commercial application of genetic engineering are considered free up to a point where the rights of others may be impinged on. Such affected rights may include consumer rights (like the right to make informed choices), rights of health protection (i.e. against human health risks associated with genetically modified products), perhaps human dignity or the right to personal identity and self-determination. However, once these concerns are met, there is nothing to stop genetic engineering from fundamentally altering the genetic structure of which nature is made up. That is why, for example, the cloning of humans may be seen as restricted by the principle of human dignity or the right to personal identity and self-determination, but the cloning of animals and plants is not. This would be purely an issue of utilitarian considerations. If the 'Dolly' experiments appear useful to humans and their immediate needs, they will be considered lawful. Sheep, like all animals and plants, are at the receiving end of our anthropocentric morality.

It may be, of course, that our morality will change over time and that, one day, ethical committees will have the wisdom and power to stop genetic engineering going mad. At the moment, ethical committees are guided by absolute freedom of research on the one hand and utilitarian cost-and-benefit analysis on the other. Since both principles are firmly enshrined in our human rights concepts, the long-term ecological implications of genetic engineering will not count.

A closer examination of current case law reveals that ecological human rights would have altered the outcome. For instance, with respect to property rights German courts have increasingly acknowledged that land and resource use is restricted by requirements of the 'public weal' (Article 14 Grundgesetz). This led, for example, to restrictions in the use of chemical fertilizers and pesticides on farmland, to protection against overgrazing caused by too many cattle, or a ban of certain hazardous substances. However, in all cases the restrictions were ultimately determined by human health standards, not ecological concerns. As the German Federal Constitutional Court (in a case in 1982 regarding ground-water levels) stated: 'Private land use is limited by the rights and interests of the general public, to have access to certain assets essential for human well-being such as water' (in Bosselmann, 1998: 115). The Court made it clear that the law cannot provide for the health of ecosystems *per se*, but only in so far as required to protect the rights of affected people. Respect for the intrinsic value of life (other than human life) would have led to much more stringent restrictions than those involved in securing water supply for people.

Judging by the ever increasing waves of the ecological movement, Catherine Redgwell is right in her observation:

> The dam of anthropocentrism has clearly been breached. Given the increasing awareness of the interconnectedness of human beings and the environment and of the intrinsic value of the latter ... nature is unlikely to simply be ignored; rather, the problem is one of reconciling a diverse environmental (agenda) and human rights agenda.
>
> (Redgwell, 1996: 73)

The reconciliation of these two agendas can be achieved by implementing the principle of the 'respect for the intrinsic value of life' into our understanding of human rights. Human rights would therefore be shaped by limitations drawn from their social *and* ecological context. Some examples – using the German fundamental rights catalogue – can illustrate the implementation of this new principle of respect for life (proposed amendments in italics; for further discussion see Bosselmann, 1998: 87–124).

Article 1 Protection of human dignity

(1) The dignity of the human being is inviolable ...

(2) The German people acknowledge inviolable and inalienable human rights *and the respect for the intrinsic value of life* as the basis of every community, of peace and justice in the world.

(3) The following basic rights shall bind the legislature, the executive and the judiciary as directly enforceable law.

Article 2 Right of liberty:

(1) Everyone shall have the right to the free development of his/her personality in so far as he/she does not violate the rights of others, *ignore the respect for the intrinsic value of life* or offend against the constitutional order or the moral code...

Article 5 Freedom of expression:

(1) Art and science, research and teaching shall be free. Freedom of teaching shall not absolve from loyalty to the constitution.

Article 14 Property:

(1) Property and the right of inheritance are guaranteed. Their content and limits shall be determined by the laws.
(2) Property imposes duties. Its use should also serve the public well-being and *the respect for the intrinsic value of life*.

The importance is not in the exact wording, but in the intention or, more precisely, the dynamics carrying the ecological interpretation of human rights.

Conclusion

Historically, the idea of human rights was shaped by two major political ideologies: first by eighteenth-century liberalism establishing the ideal of individual freedom (*liberté*) and, second, by nineteenth-century democratic principles adding the ideals of equality (*egalité*). The modern ideal of human rights conceptualized the human being as an individual in a free, democratic society.

However, it was the experience of social injustice during the nineteenth century which created the social dimension of human rights (solidarity or *fraternité*). If, in the twentieth century, we experienced environmental injustice as an additional, far-reaching threat, then it is only logical to accept the ecological dimension of human rights. Does it stretch the idea of solidarity too much to include the non-human world, or is 'respect for the intrinsic value of life' an entirely new category?

Acceptance of such a category may well be the greatest challenge for postmodern constitutionalism. There is, after all, no point in prolonging the life of *homo economicus occidentalis*. This species should vanish together with all the ideas of anthropocentrism, individualism and materialism. Postmodern constitutionalism should cater for the emerging species of *homo ecologicus universalis*.

9
Planetary Citizenship: the Definition and Defence of an Ideal

Janna Thompson

Introduction

Environmental systems and problems are no respecters of political boundaries. One of the most serious issues of our times is how ethical consciousness and political institutions can become equal to the task of solving global problems. The question this chapter takes as central is how people who belong to different societies and have different and often contrary loyalties, goals and values can become willing and able to cooperate in solving these problems. The difficulty is not merely a practical or political one – how to build institutions that can encourage or force transnational cooperation. It is also an ethical problem – a matter of identifying and justifying values, principles or ideals that favour transnational governance.

To find such ideals it is natural to appeal to the cosmopolitan tradition. This is because cosmopolitans believe that there are universal principles of right or justice, and many of them favour the development of global political institutions capable of ensuring that human rights or universal principles of justice can be realized everywhere in the world. However, cosmopolitanism has many critics. Environmentalists, in particular, have doubts about the relevance of cosmopolitan principles, and many believe that the centralized global institutions often advocated by cosmopolitans are no solution to environmental problems. These and other common criticisms of cosmopolitan principles and objectives are discussed in the first part of this chapter, 'Cosmopolitanism and its critics'.

Nevertheless, the persistence of problems requiring a transnational solution suggests that cosmopolitanism in some form needs to be revived and defended. In the second part, 'Planetary citizenship', I turn to another cosmopolitan idea – that of world, or planetary, citizenship. The purpose of the chapter is to give this notion content and to show that it can ground conceptions of value and reasons for valuing that overcome many of the difficulties encountered by traditional cosmopolitanism. My aim, in particular, is to show that it provides a promising beginning in the search for an ethical basis for transnational cooperation in the solution of environmental problems. However, common conceptions of citizenship provided by political theorists are not equal to the task of defining planetary citizenship. Most theoretical approaches to politics take as their reference point individuals conceived as autonomous actors. In part three of the chapter, 'Citizenship and cooperation', I develop an alternative approach to citizenship and the values it entails – one that requires thinking about the purposes of a political society in relation to how people conceive of themselves and their responsibilities in an 'intergenerational continuum'. I argue that individuals as participants in such a continuum are predisposed to cooperate with each other to achieve or protect the goods they value – including environmental goods. Being citizens involves sharing this responsibility and supporting or instituting forms of governance that facilitate cooperation. In the fourth part, 'Toward cosmopolitan governance', I briefly discuss why I think that this way of understanding planetary citizenship overcomes common criticisms of cosmopolitanism and helps to lay the groundwork for an ethics of global governance.

Cosmopolitanism and its critics

The present world political order is not conducive to the solution of global environmental problems. Political boundaries, and identifications and attitudes associated with them, encourage moral parochialism. Citizens of states are not generally willing to take responsibility for what happens outside their borders, and governments are reluctant to pursue policies that require sacrifice of the 'national interest' for the sake of achieving global environmental objectives. The solution to this predicament, according to some political theorists, is the creation of a cosmopolitan consciousness and a cosmopolitan political order. The assumption that the present international structure of 'national prerogatives is legitimate is a substantial obstacle in the way of global justice in the use of environmental resources', concludes Steven Luper-Foy (1995: 105–6). 'If we are so much as to survive as a species and a planet, we clearly need to think about well-being and justice internationally, and together', insists Martha Nussbaum (1990: 207–8).

Cosmopolitanism, as it is usually understood, is both an ethical and a political doctrine. It asserts the existence of universal ethical ideals or principles, and it advocates a political order in which these ideals can be universally realized. Cosmopolitan ideals and schemes have persisted through the centuries, sometimes emerging with particular vigour, as in the years after the First World War. More recently, cosmopolitan ideals of governance have become linked to the advocacy of transnational democracy (Archibugi, 1995; Held, 1995).[1] As global problems, particularly environmental problems, intensify, it is natural to suppose that cosmopolitanism should become a more popular and credible ideal. It is a perverse feature of our times that both cosmopolitan ethics and schemes of governance are widely regarded with suspicion or hostility, especially by environmentalists.

Those who doubt the existence of universal principles of justice or rights reject the ethics of cosmopolitanism. These critics point out that cosmopolitanism depends on ideals associated with the rise and dominance of Western political institutions, and they think that for this reason cosmopolitan values have no claim to be universal. Other critics complain that doctrines of universal rights or justice are too remote from the real moral motivations of individuals. The moral life of individuals, they say, centres around the communities they identify with – their nation, state, religious group, clan, tribe, or family (Sandel, 1982). But people who have such identifications, even those who acknowledge universal rights, are likely to think that they are ethically justified in giving priority to their own community and its values.

Critics of cosmopolitanism are also critics of cosmopolitan institutions. They commonly complain that centralized institutions often advocated by cosmopolitans would have detrimental effects on existing communities, cultures and values. Environmentalists, especially, doubt that cosmopolitan institutions are a solution to environmental problems – even those that are global. For example, Arran Gare (1996: 144–7) recognizes that many environmental problems are global problems and require a global orientation, but he also thinks that the environment and other things of value are under threat from globalization and the international bourgeoisie who benefit from the global economy (see also Chapter 7 of this volume). 'The struggle against global environmental problems is only likely to succeed through the development of strong nation-states committed to subordinating the operations of the market to politically defined ends – notably the conservation and preservation of the environment ... ' (Gare, 1995: 145). His reason for advocating 'a kind of nationalism' is also connected to his doubts about cosmopolitanism as an ethical ideal. Only nation-alism, he thinks, can mobilize people to bear the costs of the struggle against harmful global tendencies. Pure cosmopolitanism is too rarefied an orientation for most people (1995: 146).

Many environmentalists think that even national governments are too large and centralized, too influenced by economic criteria and unresponsive to local needs to protect environmental and human val-ues. They advocate decentralization, local autonomy, the rootedness of individuals in their local community and their identity with the partic-ular environment in which they live (Sale, 1980). Advocates of decen-tralization are not opposed to intercommunity cooperation – including cooperation at a global level. However, one problem with their posi-tion is that the local loyalties and identifications that they want to fos-ter may not be compatible with the solution to global environmental problems. The difficulty is both psychological and moral. People who identify with their nation or their local community are likely to perpet-uate the rivalries and tensions that prevent serious problems from being solved. People who think that they are justified in their loyalties are likely to regard themselves as entitled to refuse to sacrifice national or local interests for the sake of global objectives. The common prob-lems associated with localism and nationalism suggest that the cosmopolitan standpoint needs to be taken more seriously by environ-mentalists. We need, at least, a way of understanding how intercom-munity cooperation can be encouraged. What seems to be required as an ethical basis for environmental global governance is a cosmo-politanism capable of resolving the tension between the local and the global and meeting the criticisms that are often directed at cosmo-politan values and proposals.

Planetary citizenship

Criticisms of cosmopolitanism indicate that a revival of the cosmopoli-tan point of view must satisfy three requirements. There must be rea-son for believing that the ethical values it is based upon are universally acceptable. These values must be capable of motivating individuals to cooperate for the sake of solving global problems. They must be the basis for a transnational solidarity that would make individuals willing, when they have good reasons, to sacrifice personal, local and national interests for the sake of people of the world as a whole. And they must give rise to, or encourage, the development and maintenance of institutions that provide political and economic means for resolving global problems, but at the same time do not undermine the particular and local relations that people value.

One possibility for developing such a cosmopolitanism is to explore the implications of another cosmopolitan notion – world citizenship. For citizens have duties as well as rights; they are supposed to cooperate for the sake of common goods. Citizenship implies solidarity, a willingness to make sacrifices for other citizens. Moreover, citizenship requires an institutional embodiment. A citizen is a member of a political society, or at least a political society in embryo, and is supposed to take seriously the responsibilities that go with membership. So if the idea of being a world citizen can be made intelligible and attractive, then a promising moral basis for environmental global governance will be established.

World citizenship is a traditional cosmopolitan notion that some think might be revived to deal with environmental crises. Derek Heater, for example, wonders whether responsibility for the planet could be a new basis for world citizenship (Heater, 1996: 146ff). Fred Steward thinks that the recognition of our common dependence on nature requires the adoption of universal values and a global identity. 'Citizenship of planet earth embodies a new sense of the universal political subject beyond the context of the traditional nation state, and a refreshed awareness of equality in terms of our shared dependence on nature' (Steward, 1991: 74).

Is it likely that people will adopt the identity of a planetary citizen? Gare's doubts about the possibility of global solidarity have to be answered. For even if people do come to think of themselves as planetary citizens, this identity may be too weak to bear the weight of sacrifice required for saving the environment. Ulrich Preuss (1995) doubts whether 'earth citizenship' – the idea that we have moral obligations that derive from the fact that we are dependent on shared resources – is even meaningful.

> The term 'citizen' implies ... the idea of intensified duties of social solidarity and moral obligations *vis à vis* one's fellow citizens. If the moral duties are all alike, irrespective of the particular relations in which individuals are engaged, then the term 'citizen' makes no sense because it is merely a synonym for 'human being' ... Belonging to the human race does not constitute a common understanding of the meaning of rights, duties, of mutuality, promises etc., because this understanding is shaped in communities. Communities exist

only as a plurality of communities, because space and time limit the scope of possible human relations.

(Preuss, 1995: 117)

Citizenship implies, according to Preuss, membership in a particular, limited community. But it is not clear why this has to be so. Time and space limit who we can relate to even within a national community, but this does not prevent us from thinking of ourselves as fellow citizens. The fact that citizenship has always been understood as membership of a limited community does not mean that it can never become global. Nevertheless, Preuss's critique, and Gare's doubts about the efficacy of the ideal of planetary citizenship, challenge us to give content to the concept: to show how it differs from identifying an individual as a human being or ascribing to them human rights, to reveal what values it is based upon, what kind of duties and relations to others it requires, and to consider whether such an identification could mean very much in a world in which individuals have more particular attachments and loyalties.

One way to start thinking about the meaning and force of planetary citizenship is to consider how cosmopolitanism motivated by environmental concerns differs from more traditional cosmopolitan perspectives. The answer likely to be given by deep ecologists is that planetary citizenship cannot encompass merely relations of human individuals to each other. It must be based upon recognition that natural beings also deserve respect, that they too are part of our community. However, I will not develop the concept 'planetary citizenship' in this way. First of all, because of doubts about whether it is meaningful to value in a deep ecological way, and secondly, because in fact many people do not, and it is not clear what would persuade them to do so (Thompson, 1990). It therefore seems reasonable, at least as a first approximation to the definition of planetary citizenship, to consider what it means from a human-centred point of view: to focus on what relations between people it requires. Even from this more limited perspective, planetary citizenship requires radical changes to traditional conceptions of citizenship.

Traditional cosmopolitanism is based upon the moral premise that all individuals deserve respect as autonomous agents and have a moral right to freedom of conscience and association. Most contemporary cosmopolitans recognize that to solve problems of poverty and inequality it is not enough that individuals respect each other's liberty. Something must also be done about the structural and transnational causes of environmental damage and poverty, and this is likely to entail cooperation on a large scale, transfer of resources from the rich to the poor, and extensive limitations to the scope of individual freedom and the independence of communities (Beitz, 1979). The difficulty with their proposals is not merely that of determining what political institutions or international agreements can effectively cope with existing problems, but also of redefining what justice requires and what citizenship means in a world where people have to take more responsibility for the conditions of each other's lives.

One way of defining an adequate idea of cosmopolitan citizenship is to generalize from the concept of citizenship and its rights that became common within some societies in the twentieth century. T. H. Marshall (1994) argues that citizenship in Western democracies has come to mean more than the possession of civil and political rights. It encompasses social rights: the right to benefits and services that enable individuals to become or remain agents capable of autonomy. Contemporary cosmopolitans insist that social rights belong to everyone in the world and argue for a form of global governance that enables everyone to enjoy these rights (Barry, 1989). Cosmopolitan citizenship means having a right to resources and not merely a right to liberty.

Advocates of planetary citizenship are likely to think that even this demanding idea of world citizenship does not go far enough. To be able to make autonomous choices and live a good life individuals need a healthy environment. Citizenship, some say, should encompass not merely social rights but also environmental rights – the right to clean air, unpolluted water, and so forth (Skelton, 1991; Waks, 1996). Planetary citizenship could be understood as the extension of environmental, as well as social, rights to everyone on the planet.

One problem with a definition of planetary citizenship that contains an extensive list of rights is that those who doubt the universality of an ethics of rights are likely to be particularly critical of the extension of rights discourse to cover social and environmental matters. Those who think that strong cosmopolitan institutions are impractical will be sceptical about the possibility of organizing and legitimating the transfer of resources that would be necessary for a universal realization of social and environmental rights. And people who want to defend their local values will fear that global governance strong enough to guarantee social and environmental rights will undermine the independence of their communities.

Another problem is that many environmental values cannot be encompassed by an idea of citizenship that includes the right to a healthy environment. Many people don't merely want to ensure that they and their descendants will be able to survive. They also want to maintain the integrity of their environment, preserve existing species, protect wilderness and areas they regard as part of their heritage, and preserve natural aesthetic values. They would regard it as important to preserve these things even if they thought that the loss of them would not be detrimental to anyone's health. This means that advocating environmental rights is not always an appropriate way of treating our relations to nature or solving the difficulties associated with global citizenship. We have reason for looking more critically at the assumptions on which environmental rights are based: the ideas about society and value which have determined how citizenship has been understood by political theorists and philosophers of our tradition.

Citizenship and cooperation

Theories of rights conceive of individuals as autonomous actors. An idea of citizenship that includes environmental rights merely continues this line of thought. But individuals are not merely autonomous actors pursuing their own interests. Most people also regard themselves as participants in what could be called a 'generational continuum'. They see themselves as members of a family, as participants in a community or culture – and thus as the inheritors of a tradition or heritage which they expect to pass on to their successors. Most people value their heritage – even though they may object to some features of it – and see themselves as responsible for ensuring that their successors will be able to appreciate it.

These descriptions of how individuals think of themselves and what they want accounts for the way that many people value their environment. They think of it not merely as a resource required for individual well-being, but as part of a family or community, or national heritage, and they want to preserve it not merely for the sake of their own health, but to pass it on to their descendants. Their land and its particular features and species serve as a bond between the generations, an aspect of the continuity that they want to maintain. If this is the basis of many of our environmental values, then it is often not appropriate to treat environmental goods as the rights of autonomous individuals. We need a different conception of individuals and their motivations.

One of the purposes of politics, according to modern political philosophers in the Western tradition, is to enable individuals to cooperate for the sake of obtaining goods that they cannot get by themselves.

Nevertheless, cooperation has never played a large role in mainstream political theory. It is generally assumed that a political society should allow autonomous individuals to define and pursue their own good, and confine itself to protecting their rights. At the same time, most contemporary political theorists recognize that many individuals have no chance of living a good life unless the state takes responsibility for providing resources (Rawls, 1971; Kymlicka, 1995). Families and communities cannot always ensure that their children are properly educated or that their members receive appropriate health care without the intervention of the state. Many communities cannot protect their culture without its assistance. But this means that there is a problem in political ethics of justifying political measures that require individuals to contribute to the good of others. Making education and welfare into a right doesn't solve the problem. Why should autonomous individuals recognize such rights?

My suggestion is that traditional political ethics works with too narrow a conception of what individuals are and what motivates their actions. Individuals who conceive of themselves as participants in a generational continuum will be more inclined to demand or accept political means for achieving common goods. For such individuals regard themselves as having responsibilities as well as rights, and their duties are not just to presently existing members of their family or community, but also to their predecessors and descendants. When they are not able to carry out these responsibilities by themselves, they will be predisposed to look for ways in which they can cooperate with others who have similar responsibilities. They will be motivated to make use of, or bring into being, political relations of cooperation, and these relations will be the basis for their conception of citizenship. They will think of being a citizen as sharing a responsibility for achieving common objectives and maintaining a heritage to pass on to their descendants. For them citizenship involves more than a respect for the autonomy of other individuals, but an identity with past and future generations of their society, and a concern for maintaining a continuity through the generations. It encourages relations of solidarity. For people who cooperate with each other to maintain a good for themselves and their descendants are likely to appreciate their interdependence. Through the course of time these relations of interdependence may themselves generate a heritage that individuals value and want to maintain.

What I am describing is not an idealistic conception of what individuals ought to be like. Cooperation, and the solidarity it generates, has played an important role in making modern political societies possible. The idea that people of a national society belong together, that they ought to provide assistance to each other and pass on their heritage is, I suggest, an ethics based on their cooperative relations. But there is no reason to think that the boundaries of the nation state are a natural or inevitable limit to cooperation.

People who think of themselves as participants in a generational continuum want to ensure that their successors have a healthy environment and to be able to appreciate their environmental heritage. The existence of transnational environmental problems demonstrates that individuals and communities cannot now achieve this aim – even by using the political power of their state. This gives them a good reason to extend their relations of cooperation to those outside their borders, and to develop political means for making such cooperation possible.

This would not require doing away with national states (any more than solving problems that families cannot solve requires doing away with the family), but finding the right political means for promoting the type of cooperation needed for solving the particular problems that we face. Those who are attempting to find these means are planetary citizens. The fact that political means for effective collective action do not exist means that individuals cannot fully realize their role as planetary citizens, but they can aim toward this ideal and try to make it a reality.

Towards cosmopolitan governance

Planetary citizenship is the evolutionary development of an idea of citizenship that stresses intergenerational relations of individuals and the responsibilities that individuals are willing to assume because of these relationships and values. To be a planetary citizen is to be able and willing to participate in the achievement of collective environmental and other goods – whatever form of cooperation that requires. I have attempted to explain what planetary citizenship might mean, what conception of society and citizenship it depends upon and how it can evolve from more particular, limited ideas of citizenship. I have indicated how it could lead to the overcoming of political barriers between people and demands for new political institutions. It remains to explain how planetary citizenship so understood can overcome common criticisms of cosmopolitanism.

The first problem is that of justifying cosmopolitan values. Planetary citizenship depends upon the idea that individuals generally value their relationship to others in a generational continuum, and as a result they are also likely to value certain environmental goods – not just the goods that enable people to survive, but those that can be regarded as part of a heritage. There is good reason to think that most individuals do, at least sometimes, have these values. Most people are concerned about the fate of their children and grandchildren – not just their physical survival, but also their ability to enjoy their heritage. Most people care not just about their own children but about the children of their extended family, clan, community or nation. Most people desire that some of the things that they value will survive their death. There is no reason to believe that these values and desires are confined to people in Western societies. Indeed, it may be the case that valuing relations to others in a generational continuum is more widespread in the world as a whole than the valuing of individual autonomy.

A planetary citizen is someone who assumes her share of responsibility for the collective achievement of goods which she and virtually everyone else values. So understood, it is clear that 'planetary citizen' is not merely a way of identifying someone as a human being or saying that he or she has human rights. This conception of citizenship requires that individuals accept responsibilities (though what these are depends on the problem and the political context), and regard each other as fellow citizens because of shared responsibilities. An idea of citizenship based upon responsibility cannot be separated from a conception of justice or rights. Inherent in the idea of collective responsibility are ideas about the fair sharing of burdens, and there are likely to be contrary conceptions of what this means among the people of a nation, a region, or the world as a whole. But since the basic concern of everyone is the achievement of a good, conflicts between contrary conceptions of justice or rights are more likely to be resolved than are conflicts premised on the primacy of autonomous individuals or communities.

Tensions and rivalries between individuals, communities and nations clearly exist, but a conception of planetary citizenship based upon cooperation provides at least a psychological and moral basis for transcending political and ethnic divisions. Planetary citizenship is not supposed to replace membership of individuals in nations and other associations. It should be understood as a natural development from these relationships and the responsibilities which individuals possess because of them. Planetary citizens have to accept that regional or global interests will sometimes take precedence over local or national objectives. But a planetary citizen will recognize that these sacrifices are made for the sake of the values that she possesses as a member of a community or a nation.

For planetary citizenship to exist it is not necessary that there be a community of solidarity which includes everyone in the world. People can accept common responsibility for achieving the things they value without having a common identity. However, the conception of planetary citizenship that I advocate does not preclude a kind of solidarity which embraces everyone on the planet. Indeed, it shows how such a thing is possible. An association based upon the mutual acceptance and carrying out of responsibilities can create a sense of community – particularly if decisions about responsibilities are made fairly and democratically. If a community is built on such a positive basis, there seems no reason why its identity should have to depend upon the existence of outsiders or relations of conflict.

I have tried to show that the idea of planetary citizenship is meaningful, that it has an ethical foundation, and is capable of motivating widespread cooperation. Whether it will ever become a reality remains to be seen. If cooperation in solving global problems is ever achieved it will probably depend not on governments, but on citizens of states who recognize the inadequacy of their own efforts to preserve what is important to them and are willing to establish connections with people elsewhere in the world. The growth of informal transnational links between environmentally aware individuals and non-government global organizations concerned with the environment are promising beginnings. So would be the moves toward transnational democracy championed by Held and Archibugi (see Chapter 13 of this volume). The solution of global crises depends upon the growth and vitality of these developments and on widespread recognition that the environmental well-being of the planet requires a new stage in the development of political cooperation, a new social contract so to speak. While this chapter does not explain specifically how this is to be accomplished, it does point the way to a cosmopolitan ethics that can justify and motivate the search for a new form of global governance.

Note

1. I regard my attempt to define and defend planetary citizenship as complementary to Held's (1995) attempt to redefine the conception of democracy and to find an ethical basis for democratic governance that can underwrite its extension to the global order.

10
An Ecological Ethics for the Present: Three Approaches to the Central Question

James Tully

Introduction

Reflecting on the most pressing issues of our times, Arne Naess has stated that the 'central question' is, 'how can the fact of cultural and philosophical difference on justice and nature be reconciled with the urgent need to deliver fair judgements in cases of conflict between development and the environment, exploitation and conservation?' I agree that this is one of the central questions of the present. The importance of the question is that it orients critical reflection not towards some abstract question of worldviews or of an imaginary world beyond conflict, but towards what is happening here and now: to the conflicts over our relation to the environment and how they are to be addressed.

In response, I would like to sketch an ethics, a way of thinking and acting, appropriate to this situation of environmental conflict in which we are engaged. By an 'ethics' I mean an approach that enables people to analyse critically cases of environmental conflict on the one hand and to act ethically and effectively to bring about fair judgements on the other. I will begin by introducing this type of ecological ethics as a response to the limitations of two better-known alternative and complementary approaches.

The first and most prominent approach is to try to work out very general principles of environmental justice that should apply to any situation of conflict. These principles are usually articulated in terms of universal rights and duties and their global institutionalization. David Held's theory of cosmopolitan democracy is an excellent example (Held, 1995, 1998). Notwithstanding its great strengths, cosmopolitan democracy has two limitations relative to Naess's central question.

First, its perspective is the long term, not 'the urgent need to deliver fair judgements' in immediate 'cases of conflict'. Second, it does not start from the present 'cultural and philosophical difference on justice and nature'. Rather, it takes 'autonomy' to be the supreme value and derives universal environmental rights, duties and institutions from it. Democratic discussions of conflicts over the environment take place within this framework of the accepted priority of autonomy and the rights and duties derived from it.

This raises three objections. First, as Charles Taylor argues, the very idea of deriving a system of justice from a single value is dubious. There are always several values, principles and goods brought to bear by participants in a conflict, whose ordering, interpretation and application are open to disagreement and which vary to some extent from case to case (Taylor, 1994: 246–9). Second, most ecologists would not rank autonomy as highly as Held does, let alone exclusively, and thus the thought experiment by which he tries to establish it would fail. For example, Fritjof Capra suggests that the relevant ecological values are not autonomy, but 'interdependence, recycling, partnership, flexibility, diversity, and as a consequence of all those, sustainability' (Capra, 1997: 304). Third, if environmental justice is to be democratic, then the principles, values and goods that are brought to bear in a conflict must themselves be open to democratic discussion and debate. They cannot be decided monologically by a theorist, but must be agreed to by the people affected by the conflict through democratic dialogue.

The second general approach takes the third, 'democratic' objection seriously and leaves it to citizens themselves to reach agreements on the norms of environmental justice in processes and institutions of deliberation over cases of conflict. The aim of this dialogical or deliberative approach, accordingly, is to work out the conditions of fair and just deliberation among all those affected by a conflict over the environment. John Rawls's conception of an 'overlapping consensus' and Jurgen Habermas's theory of 'discourse ethics' are two well-known examples of this approach (Rawls, 1993, 1995; Habermas, 1995a,b). The basic idea is the rule of democratic legitimacy, *quod omnes tangit*, what affects all must be agreed to by all. In Habermas's formulation, the democratic principle D is that only 'those norms can claim to be valid that meet (or could meet) with the approval of all affected in their capacity as participants in a practical discourse' (1995a: 66). However, as many commentators have pointed out, there are limitations to this approach as well. First, both Rawls and Habermas screen out rather than allow for deep cultural and philosophical differences

(Tully, 1995), although a modified version of Rawls's theory has been advanced in response to this objection (Laden, 1997). Yet, as Low and Gleeson persuasively argue, any reasonable and practicable global ecological ethics and politics 'should be specifically designed to reinforce and constitute' cultural diversity (1998: 194).

A necessary condition for resolving disputes over the environment is that people's background conceptions of justice and nature are brought into the discussion and criticized through a reasoned exchange. This is, for example, the democratic way to bring about what many ecologists see as a paradigm shift from a mechanistic to an ecological view of nature (Capra, 1997). However, in Rawls's theory, and models of dispute resolution based on it, a reasonable pluralism of conceptions of nature is accepted as the given background on the basis of which citizens enter into discussions to reach an overlapping consensus on principles of justice. Although this conservative approach may occasionally yield agreements on relatively shallow conflicts, it does not enable the participants to call into question the deeply sedimented background conceptions of nature that block fundamental change. These conceptions include the dominant view that environmental damage is an externality or that nature can sustain unlimited growth.

Although Habermas's theory allows for a wider range of critical questioning, it excludes from discussion the very form of ethical reasoning that ecological conflicts require according to the vast majority of ecologists. Habermas draws a sharp distinction between dialogical moral reasoning, deontological questions of justice for each and every individual, and dialogical ethical reasoning, evaluative questions of the common good for the members of a community. He holds that only the former has the capacity for universal and unconditional agreement (Habermas, 1995a: 108; Rehg, 1994: 92–150).

For him, this differentiation of justice from ethics has to be accepted as an inescapable feature of modernization (1995a: 17–20). However, for many ecologists it is precisely the attempt to differentiate the just from the good and to treat humans as autonomous entities that is the basis of the conflict. For ethical ecologists, humans exist within, are dependent upon, and are members of the web of life, the innumerable ecosystems which make up the living world, Gaia, which westerners call the 'environment' (Capra, 1997: 3–16). As Low and Gleeson conclude, 'the relationship between humanity and nature is best described as asymmetrically co-dependent. We can appreciate today that the survival of the natural world is dependent upon what humanity does. At the same time humanity remains completely dependent for survival upon non-human nature, that is to say, upon our planetary biosphere and all its inhabitants' (1998: 155–6).

This 'ecocentric' as opposed to 'egocentric' view of humans' place in the world is the basis of the deep ecology, social ecology, ecofeminism and spiritual ecology (Merchant, 1992: 61–156). In addition, it is the emerging vision of the interconnected network of all forms of life in the life sciences and systems theories (Capra, 1997: 157–285). Moreover, as Capra points out, it accords with the cultural and spiritual wisdom of the ages:

> Ultimately, deep ecological awareness is spiritual or religious aware-ness. When the concept of the human spirit is understood as the mode of consciousness in which the individual feels a sense of belonging, of connectedness, to the cosmos as a whole, it becomes clear that ecological awareness is spiritual in its deepest essence. It is, therefore, not surprising that the emerging new vision of reality based on deep ecological awareness is consistent with the so-called perennial philosophy of spiritual traditions, whether we talk about the spirituality of Christian mystics, that of Buddhists, or the philos-ophy and cosmology underlying Native American traditions.
>
> (Capra, 1997: 7)

If a basic aspect of the human condition is an interdependent rela-tion in the environmental network or web of life, then the question arises of what ethical comportment should humans take to this rela-tionship of interdependency within the larger ecocommunities or ecosystems? The answer is that we should take up the appropriate atti-tude of care, concern, respect, responsibility and perhaps awe for the value of all living things which compose the larger web of life (Low and Gleeson, 1998: 133–55; Capra, 1997: 12). This ethical orientation to the common good of the ecocommunity will be an inseparable dimension, therefore, of any democratic discussion aimed at reaching fair judgements in conflicts over the environment.

Of course, deontological questions of justice remain, but these can-not be discussed in isolation from the ethical questions of our relation to nature (Low and Gleeson, 1998: 156–7). For example, the six 'prin-ciples of ecological sustainability' presented by Mark Diesendorf and Clive Hamilton in *Human Ecology, Human Economy* illustrate the inseparability of ethical, moral and ecological considerations: the conservation of biodiversity and ecological integrity, the conservation of cultural diversity, the improvement of individual and community well-being, intergenerational equity, the precautionary principle, and community participation in decision-making (1997: 64–98). If this analysis of humanity's relation to nature and an ethical orientation to that relation is correct, then Habermas has the relation between morality and ethics the wrong way round. It is ecological ethics that is global and universal whereas deontological morality is a limited, species-centric, or ego-centric, perspective. This reaffirmation of the priority of ethico-political reasoning is one of the many lessons of the classic text on ecological political economy by Herman Daly and John Cobb, tellingly entitled *For the Common Good* (1994).

Accordingly, any ethical approach to ecology, including the one presented here, will address the central concerns of ethics: what ethical orientation should we take to our relationship of interdependency with other members of the web or community of life (care, stewardship, respect)? What are the reasons why we should adopt this orientation (scientific, spiritual, pragmatic)? What practices of self and group formation should we engage in to constitute ourselves as fit members? What is the telos or good of this way of being in the world? (Foucault, 1985: 28–32). There is a plurality of answers to these concerns and so a plurality of ways of acting ethically in relation to the environment today, as there has been historically even within European societies (Glacken, 1967).

These considerations also lead to the reformulation of the democratic principle D. 'All affected' must include not just humans but all living things, and not just this generation but future generations. Since all cannot be actual 'participants in a practical discourse' over a case of conflict, they must be represented in some form. Hence, a realistic principle of democratic legitimacy will be a principle of representative democracy, RD, under which at least some representatives will take up the responsibility for presenting the ethical considerations of care for all members in the web of life affected by the conflict in question.

Finally, Habermas makes the highly idealized assumption that the practical discourse on how to resolve a conflict on the environment is free of relations of power. He does this in order to develop a normative form of argumentation that can be employed as a regulative ideal to judge the validity of any actual negotiation. However, rather than throwing critical light on actual cases of conflict and their resolution, this approach tends to lead to abstract and utopian theory (Blaug, 1997). If we are to develop a form of analysis that is enlightening and enabling with respect to actual cases of conflict, then it has to take the form of an immanent critique of, rather than an abstraction from, the existing relations of power in any process of democratic negotiation over an environmental conflict. I agree with Michel Foucault's objection to this feature of Habermas's discourse ethics and his alternative to it:

> The idea that there could exist a state of communication that would allow games of truth to circulate freely, without any constraints or coercive effects, seems utopian to me. This is precisely a failure to see that power relations are not something that is bad in itself, that we have to break free of. I do not think that society can exist without power relations, if by that one means the strategies by which individuals try to direct and control the conduct of others. The problem, then, is not to try to dissolve them in the utopia of completely transparent communication but to acquire the rules of law, the management techniques, and also the morality, the *ethos*, the practice of the self, that will allow us to play these games of power with as little domination as possible.
>
> (Foucault, 1997a: 298)

In summary, an appropriate ecological ethics for the conflictual world we inhabit will seek to overcome the limitations of these two better-known approaches in responding to Naess's central question. It will start from actual contests over ecologically damaging forms of conduct in relation to nature and the modes of dispute resolution that arise from them. It will accept value pluralism, cultural diversity and the principle of representative democracy, and allow for the critical discussion of cultural and philosophical differences over justice and nature, including egocentric and ecocentric orientations, in the process of reaching fair judgements. It will also analyse what lies at the centre of the conflict – the existing power relations that direct and control the disputed relations to nature – with an aim of changing these relations in accord with a more appropriate ethical mode of care for the network of living beings affected. It also will be an experimental and prudential ethos, rather than a universal solution, in the sense that critical reflection on one experiment in modifying our relation to nature will provide the basis for the next.

The main aspects of this ecological ethics are presented in the following three sections: 'practical systems', the context in which conflicts over the environment arise; 'negotiations over the central question', the ethical and strategic activity of reaching fair judgements in cases of conflict democratically; and 'implementation and review', the responsibility of critically monitoring environmental agreements and their institutionalization. A brief defence of this approach concludes the chapter.

Practical systems

The first step in this ecological ethics is to examine the practical context in which the central question arises. The context is, as we have seen, a conflict between development and the environment or exploitation and conservation. That is, the way a specific form of organized activity or practice affects the environment is called into question and challenged, and a conflict arises over how to settle it between (schematically) those who support development and those who support sustainability. Before turning to the procedures of democratic negotiation, let us analyse the form of organized activity in which we are constituted as agents acting on nature and over which the dispute irrupts. This can be anything from the recycling of wastepaper in an office to activities of resource extraction and production, any coordinated human activity that affects the environment.

I would like to adapt the form of analysis Michel Foucault and his students have developed for the critical study of conflicts or struggles to challenge and modify various practices of human activity. He calls the organized forms of human activity he studied throughout his career 'practical systems' and analyses them in the following way:

> Here we are taking as a homogeneous domain of reference not the representations that men give of themselves, not the conditions that determine them without their knowledge, but rather what they do and the way they do it. That is, the forms of rationality that organise their ways of doing things (this might be called the technological aspect) and the freedom with which they act within these practical systems, reacting to what others do, modifying the rules of the game, up to a certain point (this might be called the strategic side of these practices). The homogeneity of these historico-critical analyses is thus ensured by this realm of practices, with their technological side and their strategic side.
>
> (Foucault, 1997b: 317–18)

These practical systems stem from three broad areas: relations of control over things, relations of actions upon others, relations with oneself. This does not mean that each of these three areas is completely foreign to the others. It is well known that control over things is mediated by relations with others; and relations with others in turn always entail relations with oneself, and vice versa. But we have three axes whose specificity and whose interconnections have to be analysed: the axis of knowledge, the axis of power, the axis of ethics. In other words, the historical ontology of ourselves must answer an open series of questions; it must make an indefinite number of inquiries which may be multiplied and specified as much as we like, but which will all address the questions systematized as follows: How are we constituted as subjects of our own knowledge? How are we constituted as subjects who exercise or submit to power relations? How are we constituted as moral subjects of our own actions?

The 'technological' side of a practical system is the context for ecological ethics and the 'strategic' side is the actual ethical activity in this context. Turning to the analysis of the four areas of the 'technological' aspect, a practical system will involve, first, relations of production and distribution which affect the environment. The participants in the system (such as workers, managers, distributors, consumers, investors) will be constrained to act in accordance with, to sustain and to develop these productive and distributive relations, or, from an ecological perspective, 'relations to the environment', as we can call them (roughly Foucault's 'relations of control over things').

Second, the conduct of the humans engaged in sustaining and developing these relations to the environment will be directed and controlled by two types of power relations, or, as Foucault calls them elsewhere, relations of 'governmentality' or simply 'government' in the broad sense of any mode of guiding the thought and action of others in a relatively stable and predictable way (Foucault, 1984: 216–26; Dean, 1999). Their conduct will be governed more or less by the regulations and laws of the firm, sector, union, municipality, provincial and regional governments, nation state, NAFTA, GATT, WTO and international laws, treaties and agreements of various kinds. There will also be a complex system of governance, rarely democratic, of the specific practical system itself, involving the coordinated interaction of workers, unions, technicians, managers, directors, chief executive officers, shareholders, reflexive monitors and so on, through which the regulations and laws are operationalized or evaded (Foucault, 1984: 217–19). Foucault's 'relations of action upon others' refers to both these types of power relations and the ways in which they constitute the forms of subjectivity of the practitioners of the system (that is, the mental and behavioural competencies characteristic of their roles).

Third, members of a practical system will have distinctive ways of thinking and acting within the broad relations of governance of the system (what Foucault refers to as 'relations with oneself' or 'the axis of ethics'). Each of these three 'areas' involves forms of knowledge (scientific, technical, managerial, environmental, regulatory, administrative, economic, legal, political and psychological disciplines) that are employed in production and distribution, and in the governance of men and women.

Fourth, as Foucault mentions in another text, a practical system also involves 'relations of communication' through which the agents involved coordinate their various activities (1984: 216–19).

As Mitchell Dean summarizes, the analysis of the 'forms of rationality' of a practical system is a study of a system of governmentality:

> Government [or governmentality] is any more or less calculated and rational activity, undertaken by a multiplicity of authorities and agencies, employing a variety of techniques and forms of knowledge, that seeks to shape conduct by working through our desires, aspirations, interests and beliefs, for definite but shifting ends and has a diverse set of relatively unpredictable consequences, effects and outcomes. An analysis of government, then, is concerned with the means of calculation, both qualitative and quantitative, the type of governing authority or agency, the forms of knowledge, technique and other means employed, the entity to be governed and how it is conceived, the ends sought and with the outcomes and consequences.
>
> (1999: 42)

Studies of four 'areas' of practical systems and their interconnections will be necessarily wide-ranging: examining both the local context and its global connections, what is happening right now and critical histories of the formation of the specific relations to nature (Darier, 1998; Rutherford, 1998).

If the technological aspect of practical systems is the context in which humans are constituted as subjects acting on nature in a certain way, then the 'strategic side of the practices' is the freedom they have to call the practice into question, to enter into some form of 'conflict', and to seek to 'modify the rules of the game up to a certain point'. Any relation of power, no matter how strictly enforced, involves the possibility of freedom on the part of those over whom it is exercised.

Power is exercised only over free subjects, and only insofar as they are free. By this we mean individual or collective subjects who are faced with a field of possibilities in which several ways of behaving, several reactions and diverse comportments may be realized.

(Foucault, 1984: 221)

Hence, a case of conflict over some area of the technological organization of a practical system is, in Foucault's terms, the strategic exercise of freedom by the members who call into question and challenge the prevailing 'rules of the game'. He characterizes the permanent relation between power and freedom as an 'agonism', the permanent possibility of contesting, rather than acting in accord with, a relation of power:

At the very heart of the power relationship, and constantly provoking it, are the recalcitrance of the will and the intransigence of freedom. Rather than speaking of an essential freedom, it would be better to speak of an 'agonism' – of a relationship which is at the same time reciprocal incitation and struggle; less of a face-to-face confrontation which paralyzes both sides than a permanent provocation.

(1984: 221–2)

An environmental conflict erupts, therefore, when some members of a practical system call into question and contest the degrading relation to the environment that their organized activities sustain and develop. Applying the democratic principle that 'what affects all must be approved by all', any living member of the ecological web of life affected by the productive or distributive activities of a practical system, or their representative, may be considered a member with the democratic right to call the activities into question. Since the ecological effects of a local system are often global, there is a 'disjuncture', as Held puts it, between the traditional understanding of a democratic community bounded by a territorial nation state and the global reach of environmental degradation (Held, 1995, 1998: 19–20). For example, consumers or members of Greenpeace in distant countries organize democratically to boycott the products of Canadian forest companies in order to contest and modify their environmentally damaging forest practices, in concert with environmental activities at the local sites of forestry.

The emergence of local conflicts and democratic action throughout the affected network of life, independent of the country of origin of the environmental problem, may lead in the long run to the extension of the traditional legal and political institutions of western representative democracy to the global level, as cosmopolitan democrats hope. However, this is not the case at present. The emerging global institutions form an unacceptable and ineffective 'negotiated order' of global governance (Low and Gleeson, 1998: 175–83), and the general trend of the institutions, such as NAFTA and WTO, is in the opposite direction: to disempower local democracies and deregulate production and distribution (Nader and Wallach, 1996). Rather, these forms of conflict, organized strategically in immediate response to the specific area of the practical system in which they are engaged, and coordinated with other site-specific struggles throughout the affected network of life, should be seen and studied carefully in their own right, as a quite distinct form of the 'relocalization' or 'eco-networking' of democratic ecological politics (Mander and Goldsmith, 1996: 393–514; Foucault, 1984: 211–12).

To call a specific relation to the environment into question, one needs, of course, to show how it adversely affects the environment, to be able to challenge the validity of the scientific knowledge that is employed to legitimate the practical system in its present form. It needs also to be shown that the relation could be otherwise, that this activity, or a suitable substitute, could be organized in a way that cares for and sustains the environment. It is easy enough to state this in general terms, as Herman Daly points out (1996: 195–6), but, to be convincing in the negotiations leading to 'fair judgements' the case must always be made specifically; that is, for this locale and this ecoregion (Sale, 1996).

These changes in turn almost always entail modification in the four areas of the technological side of the practical system, even in the relations of communication, say, if only to get the proposed changes on the agenda. In order to be convincing, one needs to show that the prevailing form of organization is historically contingent and could be otherwise, and indeed to be able to present an alternative that is not only ecologically sound but also addressed the legitimate concerns of those affected by the changes (workers, local groups, consumers, and so on). Moreover, one needs knowledge of early attempts, or similar attempts elsewhere, to change the system in question, so the strategic activity is not needlessly ineffective, easily co-opted or reinventng the wheel. This is why detailed studies of the technological side of the practical system are indispensable to an effective ecological ethics on the strategic side.

Negotiations over the central question

Let us now imagine that some members of a practical system have been able to call its relation to the environment into question and those who exercise power have been constrained to respond by entering into negotiations of some kind in order to resolve the conflict. That is, it is a case of conflict in which Naess's central question arises. As we have seen, such a conflict can occur anywhere, from the traditional legal and political institutions of representative democracies, to negotiations over environmental amendments to NAFTA, to site-specific struggles of local democracy in the workplace, dumpsite, forests, stores and so on. More often than not, they occur across the two types of power relations that govern systems of production and distribution mentioned above: the negotiated order of local, regional, national and international governance and the forms of governmentality of the practical system in question. Members of a practical system demand a say in the way in which they are directed to act on the environment, thereby democratizing the system to that extent, and they coordinate these activities with traditional political and legal action.

The importance of the local action in the first instance is that it challenges an organized form of activity from the inside. Recall that a practical system constitutes to a considerable extent the characteristic ways of thinking and acting in regard to the environment of its members, that is, their forms of self-awareness and self-formation. The fact that they are able to challenge these powerful processes of subjectification shows that we are not completely determined by the systems in which we are engaged. Further, the challenge is not vague or abstract, unrelated to concrete practice, but a specific way of thinking and acting differently that emerges in the context of, and in an agonic relation to, the sedimented structures of the environmentally degrading activity it problematizes. In this way, the practice of ecological ethics takes place on the ground of practical systems and the strategies of freedom to modify them from within.

In the first section, I enumerated the sorts of normative conditions that render such negotiations legitimate. The people and other living 'stakeholders' in the ecosystem affected by the contested form of activity, and so having a say directly or through representatives, will be various, they will have a variety of different concerns, and a variety of cultural and philosophical ways of seeing the situation and presenting their pros and cons (Young, 1996). The studies of the technological aspect of the practical system will equip ecologists to enter into the various specialized negotiations over the scientific, economic, legal and political ramifications of the current system and the proposed changes. In addition, the studies of the relations of power prepare participants to respond to fake negotiations, backsliding, bribes, threats to close and relocate, and the like, as well as to organize their own networks of support both locally and globally.

One of the most difficult exercises in negotiating fair judgements is bringing the other side around from a perspective of unlimited growth and development to see the situation from an ecological point of view: that is, from our interdependency in the web of life and an ethical stance of care. The forms of practical reasoning in negotiations with others who have different perspectives are much more complex than simply taking a 'yes' or 'no' position on a proposed norm of action-coordination and giving one's reasons. Those who wish to introduce an ecological orientation of sustainability need to show how it relates to the values, principle and goods the others already hold as well as showing how it answers their legitimate concerns about employment, efficiency, future generations and a host of other considerations. Furthermore, these discussions often take place across cultural, philo-sophical, class, gender, age, regional and other differences on the central issues.

At the heart of ecological ethics, therefore, is the principle of *audi alteram partem*: always listen to the other side. This is not simply a duty of respect to differently situated others who have an equally legitimate right to speak and be listened to. The way in which we listen to others who have different points of view, enter into dialogue in order to try to understand where they are coming from, then try to respond in a way that enables them to understand our point of view is also how we free ourselves from our own sedimented self-understanding of our relation to the environment. It is also how we see the limited and partial char-acter of this self-understanding, and, with the help of the dialogue with others, begin to move around to a broader view the relevant con-siderations, and so open the possibility of reaching a fair judgement. The negotiations are not simply a site of strategic bargaining. They are the intersubjective dialogues in which we come to acquire and appreci-ate the cultural and biological diversity of our interdependent relation-ship to all relevant aspects of the web of life (Tully, 1995: 183–212). This is a form of self-awareness and self-consciousness ('diversity' or 'aspect' awareness) appropriate to an ecological ethics.

Two examples will illustrate this point. As Naess suggests, negotia-tions are often characterized in such broad oppositions as 'development versus the environment' and 'exploitation versus conservation'. One of the advantages of entering into complex negotiations with the kind of studies outlined above is that these broad oppositions tend to break down in the course of the discussions. The public negotiations over the forest system on the North-West coast of North America have been structured for many years around development versus care for the environment: development and employment were claimed to be dependent on respecting the imperatives of capitalist growth and the globalization of the economy.

By studying the local economy and the forest industry and its relations to the global economy Michael M'Gonigle (1994), Patricia Marchak (1995) and Jeremy Wilson (1998) have shown that this is a misleading and disempowering way to structure the debate. They argue that present forest practices of extraction and export are not only environmentally destructive; they also lead to a decrease in employment and the destruction of local communities and economies. Alternatively, M'Gonigle in particular has argued convincingly that ecologically sound forest practices organized on local and regional bases are compatible with, and the means to an increase in, local employment and a more diverse and self-reliant economy. Thus, by a careful analysis of the practical system and the possibility of modifying it, the way the discussions have been structured by the powers-that-be has been changed and those concerned about jobs have been brought around to see their concern addressed persuasively from an ecological perspective.

The second example is negotiations involving land use with indigenous peoples. As is now well known, indigenous peoples bring to the negotiations quite distinctive practices in relation to nature (Yunupingu, 1997). If non-indigenous people are to understand what they are saying and to learn from them, then they need to be able to free themselves from their own unreflective understanding of the environment and their relation to it, whatever it may be (Knudtson and Suzuki, 1992). They can do this, as far as I know, only through the kind of critical dialogue of reciprocal elucidation sketched above, such as the 'two-way or *ganma* dialogue' developed by the Yolngu people of Arnhemland (Watson and the Yolngu community at Yirrkala, 1989).

The objective of these discussions is not to exchange Western and indigenous worldviews on the environment, but to understand the different practices in which Western environmental knowledge and the traditional ecological knowledge of indigenous peoples are embodied. The specific ecological practices of indigenous peoples are analogous to the practices by which non-indigenous peoples are led in their practical systems to recognize themselves under a specific relation to the environment. I do not see how we can understand what indigenous peoples are trying to say about alternative ecological practices unless we have, by means of the analyses outlined above, grasped the ecological practices in which our own thought and action is shaped and formed. It puts us in a position to enter into a dialogue with them, to come around to see our relation to nature from their point of view, and so to begin the important and indispensable exercise of learning comparatively from each others' ecological practices through cross-cultural dialogue.

Historical studies of the formation of Western practices can also play an important critical role in these dialogues. In societies founded on the internal colonization of indigenous peoples, such as Canada, Australia, New Zealand and the United States, the destruction of the diversity of indigenous cultures and the imposition of a dominant settler culture has gone along historically with the destruction of indigenous biodiversity and the implantation of an 'imperial ecology' (Crosby, 1986). These two processes of subduing indigenous peoples and indigenous biodiversity, the strategies of freedom indigenous and non-indigenous peoples have exercised in resistance to them, and the long and uneven interaction between them are beginning to be studied historically. These historical studies enable us to think critically about our relation to nature in the present by showing that our current practices are neither necessary nor universal, but historically contingent and capable of being otherwise (Merchant, 1989; Cronon, 1983).

Finally, any judgement reached by the negotiators, no matter how fair, will never be the definitive resolution of the central question or a consensus (Tully, 1999: 138–40). There are several reasons for this. Asymmetries of power, knowledge, influence and resources will play their role in any negotiations. Real time constraints entail that a judgement will be made before all those affected have been heard or have reached agreement. Unanticipated consequences in the implementation of the agreement may show that those who dissented were right after all. Moreover, in any complex discussion and agreement there is always room for reasonable disagreement (Rawls, 1993: 56–8). As Foucault puts this rather obvious but often overlooked factor of indeterminacy (1997a: 297):

> With regard to the multiple games of truth, one can see that ever since the Greeks our society has been marked by the lack of a precise and peremptory definition of the games of truth which are permitted to the exclusion of all others. In a given game of truth, it is

always possible to discover something different and to more or less modify this or that rule, and sometimes even the entire game of truth.

The indeterminacy in games of truth holds in the specific case of the human and natural sciences of the environment as well (this is the example Foucault uses to illustrate his general point). It follows that in any agreement we reach on procedures, principles, ethics, scientific studies or policies with respect to the environment, including any ecological paradigm, there will always be an element of reasonable disagreement, and thus the possibility of raising a reasonable doubt and dissension. Any judgement, whether global or local, will be a negotiated accommodation or reasonable compromise involving an element of non-consensus, not a definitive and peremptory consensus on the environment and our relation to it. Consequently, the agreement and its institutional implementation must themselves be seen as experimental and provisional: that is, open to review, question and challenge.

Implementation and review

This form of ecological ethics, then, directs critical attention to the institutions of implementation of agreements on the environment just as much as to the principles and procedures for reaching agreements. There are two reasons for this. The first, as we have just seen, is the imperfection of any agreement. No agreement will be the definitive resolution of the central question in a case of conflict. There will thus always be the need to review and call into question its implementation, and so to begin all over again.

The second reason is that the agents come to the negotiations with an understanding of the practical system out of which the conflict has arisen and in which the agreement has to be implemented. So, their ethical concern will be to link, as closely as possible, the considerations of environmental principles, global ethics and policies embodied in the agreement with its institutionalization and application in practice, where it modifies the way they act on the environment. Yet there is a tenuous connection between local agreements, charters of environmental rights and duties, precautionary principles, laws and regulations on one side, and their interpretation and application in day-to-day and year-to-year practice on the other. The ways any rule can be said to follow are wide and divergent, not to mention the ability of powerful parties to drag their feet or dissimulate over compliance. I am not saying that agreements and institutions across the negotiated order of governance from the local to the global are not important: quite the opposite. Because there is neither a definitive form of them nor a self-guaranteeing mode of implementation, they are too important to be left beyond the bounds of ecological theory and practice, as if they were some sort of separate and merely supplementary field. Ecological ethics does not come to rest with an agreement or an institution. It is a permanent task.

Conclusion

In conclusion, I would like to respond to one objection to the form of ecological ethics I have outlined. As we have seen, this form of ethics is a response to the limitations of two better-known and more universal environmental approaches, cosmopolitan democracy and discourse ethics. In contrast to them, it tends to concentrate on the present and to ground ethical activity in local practices, networked with other similar activities, and contemplates an ecologically sound global network of institutions and practices developing in due course and in unpredictable forms on this firm foundation. It is an ethics of thinking globally and acting locally. The objection is that there is a danger of being overwhelmed by global processes circumventing and beyond the control of these fragile specific activities and their ad hoc networks.

One form of this objection is that the global system of capitalist production, distribution and finance is invulnerable to these specific struggles. Such an economic system requires a global and systemic response to counterbalance its environmentally damaging effects. In reply, I find this way of thinking about global capitalism misleading and disempowering. It is misleading because the capitalist economy is not a closed, well-defined, self-steering and boundary-maintaining system in the required sense implied by defenders of global capitalism on one side and ecosocialists on the other. Rather, as Theodore Schatzki argues, it is a complex network, or constellation of networks, of overlapping and criss-crossing heterogeneous practical systems (1996: 221–5). The systemic characterization of global capitalism is disempowering because it makes it appear that ecologically concerned citizens are powerless to act where they live and work. If, conversely, the global system is a congeries of practical systems, then the most effective place to act is in the practical systems in which we find ourselves.

The second response is that this form of ecological ethics also looks forward to the long term, by promoting here and now the form of activity that will be constitutive of any ecological future. It consists in local democratic activity and this democratic activity is oriented towards making the local, bioregional practices more ecologically benign and economically self-reliant. In this respect, it is in opposition to economic globalization and the attempts to regulate the environment through global agencies, for, as was mentioned earlier, these tend to be ineffective and anti-democratic (Mander and Goldsmith, 1996).

The third response is that large-scale change in processes, structures and sedimented forms of thought that adversely affect the environment is brought about, I believe, by changing the practices in which they are embedded and reproduced. It is our routine acting that holds these seemingly autonomous systems in place. Acting differently, exercising our freedom here and now, can change them. This has been the teaching of the great philosophers of practice from Marx to Wittgenstein and Foucault, and I see no good reason to doubt it.

Finally, this practice-based approach has the potential of bringing together four areas of ecology that are currently fragmented: multidisciplinary studies of the practical systems in which we are constituted as subjects acting on nature; the local struggles and ecological movements that call these into question in practice; normative and empirical studies of the proliferating institutions and procedures of dispute resolution of conflicts over the environment; and critical, reflexive monitoring of the implementation of environmental accords. Analogous to the way Marx sought to reorient political economy around practical struggles over the relations of production in his day, this ethics could reorient political ecology around practical struggles over relations to the environment in our day.

Even if these responses have some validity, the risk of being determined by processes beyond our control remains. Our critical investigations and ethical activities are always limited, partly determined, humble, less than we hoped for. But this just means, as Foucault said in response to a similar objection, 'we are always in the position of beginning again' (1997b: 317).

11
Environmental Ethics and the Obsolescence of Existing Political Institutions

Peter Laslett

Introduction

It is easy in expounding this subject either to record what is crassly obvious or to be unrealistic and alarmist. In the situation in which we find ourselves there is however some virtue in setting out what is well known and widely agreed. An orderly recension of the information that we have and some suggestions about what we can and ought to do about it could help. But scare tactics and exaggeration, that tendency to linear extrapolation from a few not very well established facts, which has too often disfigured environmental discussion and weakened its effectiveness, will be avoided here as far as possible. We must also try to resist the tendency to excessive abstraction that so readily attaches itself to discussions of time, and environmental justice is inevitably justice over time.

I shall proceed, therefore, by laying down some straightforward propositions. The first is that justice between humans in respect of the environment, and justice also between humans and non-human animals, or between ourselves and inanimate 'nature', however those tricky propositions are resolved, does require effective institutions. Implementation has to take place, and weakness and inappropriateness in the instruments we have for the purpose must concern us.

The second proposition is that such institutions as we have show signs of losing their power to do what is wanted for environmental justice, but signs also of uncertainty of scale. They vary between the colossal and the miniature, between more than a billion in China and tens of thousands in Iceland. They are no firm presence on the global scene.

The third proposition is that existent institutions are not only inadequate to some degree but they are also inappropriate. They were not set up and developed with environmental justice in mind, for that concept is new in political, intellectual and cultural history.

Fourth comes the fact that goes some way to explain the inappropriateness. Issues of environmental justice affect all humans, and all the other entities mentioned above and are not confined to collections of them or collections of collections, that is, nations and the United Nations for the most part. Moreover the effects are not for now or for the foreseeable future. They are for ever. Existent institutions were not designed for eternity and must always be to some degree defeated by environmental challenge.

Proposition one

Although we have resolved to stick to what is obvious, it is already evident that there are complications, and that an enquiry about obsolescence in institutions raises questions about their basic character, their status and the thinking behind them. To my mind this is one of the fascinations of the new intellectual world of environmental discussion: it leads so easily to the posing of fundamental questions. For the first of my straightforward propositions – that we must have authoritative institutions – that which may seem to be quite obviously true can be denied. It has indeed been denied, in a tradition of Western political theory with a long history, that is to say anarchism.

Quite apart from the narrow autarchy of nationalism which informs every contemporary sovereign state, it is government itself, say the anarchists, that limits freedom. Freedom for humans at large to fashion and maintain ethically correct relations between men and women in respect of the environment is certainly subject to such limitations since it may infringe national sovereignty. Sovereignty in the established view is necessary to governance, which is indispensable in its turn to ordered social life. But governance in the anarchist system can and should be maintained by spontaneous collaboration between individuals. Constituted government such as we are familiar with is therefore otiose.

These two last positions I myself reject and they would I think be rejected by most concerned people. Whatever we may think about sovereignty, reliance on spontaneous collaboration offends against the conviction, which surely we all share, that we want control and change in the right direction in environmental matters, such change as we can get in an imperfect world, and we cannot wait for the creation of an ideal world order of the anarchist or any other kind. If the nation-state of which we happen to be citizens is increasingly incompetent in these respects, then it is our duty as citizens to do what we can about it, by personal conviction, by persuasion and by political action at local, national and international levels.

But it is a sad fact that most environmentally conscious people fail to take such action, to which the relative weakness so far of Green political parties, except perhaps in Germany, bears witness. This laxness in our attitudes and action gives too much room for the ecological fanatics to dwell upon their alarmist exaggerations.

We need not go further into the content of anarchism or its place in political thinking, although its principles have an unexpected relevance to the environmental movement, and are not far from those which inform the attitude which I shall end by recommending. This is particularly so of the original English anarchist writers and of the tradition which succeeded until the French, the Germans and other continental peoples took up the strain 150 years ago and bombs began to be thrown. Such actions were and are entirely alien to anarchist thinking and practice, and this disastrous association gets in the way of their being taken seriously. In propounding a radically critical attitude to established political organizations, to parties as well as to governments and nationalisms, we are harking much further back to William Godwin or even to Gerard Winstanley the Leveller.

Proposition two

Let us turn to the second of our self-evident propositions about the obsolescence of established political institutions, that they are increasingly ineffective in carrying out what we expect them to be able to do. Put bluntly in familiar terms, nation-states and the United Nations with its agencies do not and perhaps cannot control the major polluters of the environment, which are the multinational corporations. In an age of worldwide market capitalism, regulative institutions are not only in a weak position to control such activities, but as we all know are liable to be manipulated by the multinationals themselves. Some of these are more powerful than most of the world's nation-states, and collectively the corporations might well be able to frustrate the United Nations, by evasive strategy and clever propaganda perhaps, rather than by outright defiance.

This is very familiar ground, familiar even in the media in every country, always eager to prophesy doom given the opportunity, defying if necessary the capacity of multinationals to dictate what should and should not be published. More interesting, and the proper theme of this present chapter, is why these failures have come about, and especially whether this is because of the character, structure and history of the established institutions, which have made it all inevitable, rather than its being the outcome of particularly unfavourable events and circumstances. But before we get on to this central ground there are a number of significant things to be said.

It is not true of course that all environmental outrages and all assaults on our ethical relationships with 'nature' have been made by supranational corporations. Some of them have been committed by established governments, but it is also untrue that the governments of nation-states have been unwilling to set out to control the activities of those who menace the environment, or that their efforts and those of the agencies of the United Nations have not had beneficial effects. Nor can it accurately be said that those who control mining, forest clearance, manufacture, construction, the laying out of highways, the production of energy, chemical processes, biotechnology including genetic modification, and so on and so on, are themselves necessarily and personally indifferent to the impact of what they are doing on our common environment.

In the case, for example, of the 42 transnationals that came together in 1997 to declare their intention to protect the environmental future, we should take them seriously, and assume their sincerity. We should do this because not only is it unsympathetic and lacking in trust to refuse, but it is also unrealistic to suppose without convincing evidence that they must be insincere, that there are no people of good will and environmental sensibility running these corporations and signing the declaration, well aware as they must be of its potential effect on their corporate interests. If in the last resort we do have to rely, as anarchism supposes, on the good will of the next individual to collaborate with us in our intentions for the world, then we should be prepared initially to trust everyone and believe that they mean what they say. You will note that I say initially.

To do otherwise might look paranoid, a move in the direction which, as I believe, has made environmentalism less effective as a movement than it might have become. It begins to savour of the attitude of authoritarian communist regimes and their propaganda, for which incidentally even the slightest hint of anarchism was and is anathema. It could be suggested that a similar strain comes out in references to global capitalism and the global market, so frequently denounced as the enemies of environmental justice because propelled by persons wholly insensitive to environmental values since limitless acquisitiveness is their one and only motive. The assumption seems to be that capitalism will have to go forthwith, along with the market and even social development, before the field can be cleared for a saner political order. The motor car itself might not escape the righteous holocaust. Reality in the world of politics, society and beliefs has never been as simple as this, and it is not to be wondered at that those who have assumed it was so, have become labelled as 'econuts'. The inappropriateness and obsolescence of contemporary political institutions for our purposes will have to be met with realistic appraisal, plans and policy, ones which are more in accord with the preferences and common sense of ordinary people.

Proposition three

There are further considerations of this kind, to do with the frequent appeal to a traditional past when humans and 'nature' were in harmony and justice prevailed between generations. No historical sociologist concerned with traditional societies, certainly not one like myself whose interest has been in that which once existed in pre-industrial Western Europe (Laslett, 1984; Laslett and Fishkin, 1992), supposes that such societies had in place a respect for the environment or for 'nature' as we conceive of them, least of all a policy of environmental ethics.

The inhabitants of the European West, before development appeared among them, certainly had futurity in mind in some of the things they did and thought. This comes out in their awareness of the interest of posterity in the physical structures they erected: the cathedrals, the bridges, the layout of the cities and so on. They were confident, moreover, that the social entities which they created had a rationale of endurance, as it might be called, in that their full significance and usefulness would only become apparent over time. But in spite of the fact that they must have recognized that their predecessors had exhausted the deposits of the precious metals that had once existed on their continent, a palpable loss that was remedied by colonial conquest and by commercial exchange with the rest of the world, they do not appear to have had much notion of any final limitation on natural resources. They can be seen to have possessed what their representatives surviving in our time still display, a symbiotic strain in their outlook and behaviour, living as they did and do in a given set of surroundings on which they have compulsorily relied for food, shelter and the means of living itself.

It is also true that they had a strong sense of their own indebtedness to their predecessors (Laslett, 1992). However, I have found no evidence that they felt obliged to repay the debt or at least refrain from impairing the natural inheritance of their successors.[1] In fact it is very uncertain whether they were aware of the possibility of damaging the natural world as a whole in any way whatever. Their agricultural practices were decidedly exploitative rather than preservationist. By the eighteenth century in Europe the intellectuals looked upon wilderness with horror and its clearance with satisfaction.

That we can learn from the outlook of persons who live and lived in what we are pleased to call 'undeveloped' or even 'savage' and 'barbarian' societies is no doubt true. That we should most decidedly do nothing to interfere with the natural resources on which such societies continue to rely is granted by everyone responsive to the issues. Putting an end to those activities of the northern, industrial world that go in this direction, is one of the things that we ourselves expect as a matter of course of contemporary constituted authority, national and international.

The possible effect of the obsolescence of that authority weakening such controls is a capital point in the discussion. But the discourse of environmentalists has not ended here. For it frequently seems to dwell on the imperative of preserving traditional social structures for their own sakes and for maintaining that variety in the social world which the preservation of species represents in the natural world. On this view, we as twenty-first-century Westerners should contemplate whether we ought not to abandon our high industrial condition and live as our predecessors did in the medieval era – attempt in fact to restore what I have called 'the world we have lost' (Laslett, 1983).

To go to any length in these directions is in my view entirely to lose touch with the realities we are faced with, and indeed to essay an impossible task which is not our proper concern. It is such unfortunate tendencies that encourage the harder headed and better informed commentators on the ethical problems of environmental justice to publish works with such titles as *Small is Stupid. Blowing the Whistle on the Greens* (Beckerman, 1995).[2]

Let it be added that meeting the challenge of a book like this, indeed all discussion which enlightens and informs environmentalists, has crucially important purposes. The convictions of knowledgeable people at large have so far been the major incitement to remedial environmental action and we must not let any doubt we may have concerning the effectiveness of the agencies that have taken that action, and are continuing to take it, interfere with our confidence in the value of the spread of knowledge and of the intensification of our reformist attitude. This is, after all, the world of politics, persuasion, power and even propaganda, but propaganda which has to stop short of unrealistic exaggeration and distortion.

Proposition four

Let us turn at last to what I believe to be the objective reasons for the tendency towards obsolescence in extant political institutions in relation to what is required to establish and maintain a proper environmental ethic. We begin with that supposedly omnicompetent political instrument, the nation-state, its government with its multifarious offices and subordinate local institutions. The inadequacy for environmental purposes of such an entity, or of such a complex of entities, stands out at once because they differ so much in size and in efficiency. What is more, in such matters provision has to be made for persons, indeed all persons, living outside national borders.

There is little point, therefore, in just adding environmental security and control to the list of traditional functions of national and local government, alongside external defence, internal safety and tranquillity, welfare, education and so on. Environmental ends can only be assured to a national population or any part of it if its government negotiates and consistently maintains agreements with other governments for the purpose, and we have seen that it is all other governments which have to be in question because environmental damage can be brought about virtually anywhere and environmental liability affects every single citizen of every single state in the world, along with all other humans who do not belong to nation-states at all.

Looked at from the global point of view, and this is the only one open to an environmentalist, it is difficult to imagine a more rickety edifice of authority than this, a worldwide network of agreements between governments – not peoples but their governments – each one of which has as its overriding function the defence of national sovereignty, no matter what infringes it, the environment or anything else. 'Assured' was too strong a word to use in the previous paragraph, for it is evident that the protection either of the citizen or of 'nature' must always be to some extent partial, limited to some degree in space and in efficiency, and to the greatest possible degree, limited in time.

I do not have to dwell for very long on the constitution and the history of the United Nations as an assemblage of nation-states and its agencies along with their performance in relation to the environment to drive this point home. All that is necessary is to point to the agreements made at Rio and at Tokyo with their aftermaths. The situation of populations not organized into nation-states has, however, to be noticed. The universal tendency here is to blandly assume that although such populations are small and undeveloped, it won't be long before they too become nation-states or parts of nation-states. This may be so, though not necessarily, and the really significant thing about such a comment is that the nation-state is so much the medium of all political perception that it is seemingly impossible at present to conceive of political organization in any other way. A worldwide organization and an executive which would take into account the environmental needs of every single extant human looks to be a very long way off indeed.

Sweeping as these statements have to be, they are only the beginning of the shortcomings of nation-states, or any association of them, as instruments of environmental regulation. To their limitations in respect of space has to be added their limitation in respect of duration. As has been said we have to reckon not in years, decades or centuries, but in thousands or millions of years, or indeed for indefinite time. This is a scale beyond the reach of nation-states even though each one of them presents itself as being very, very old, with a very, very long future ahead of it. Not all that convincing, from our point of view, this familiar claim. Where now is the great Soviet Union as it was between the 1940s and the 1980s? Or the worldwide polity of the British Empire which only 50 years ago boasted that it contained a sixth of the world's peoples? A community of citizens relying on either of these for environmental purposes has now to look elsewhere. And yet both these national political institutions had more political authority than the United Nations, one of whose functions could perhaps be claimed nevertheless as the provision of that permanence which the nation-state cannot guarantee.

In practice, as we have seen, environmental control through the global political process in the contemporary world does not, cannot, rely on the United Nations. It has to be backed up by concerted agreement and subsequent action on the part of governments of nation-states, as when over a hundred national governments were reported in

1977 to have agreed to tighten control over chemicals whose emissions, scientists warn us, threaten the earth's protective ozone layer. But 100 governments is not the whole world: the agreement falls short of a total ban: the most powerful nation-state, the USA, opposed the move because 'the threat does not warrant the cost to business'. Who said so? Whose business? What are the figures? Why were the criticisms of the Friends of the Earth overridden? Opposition by an individual nation-state encourages resistance by others, especially when the objector is strong and influential. This example of itself makes plain the uneasy position of such regulations, agreed between the powers, their susceptibility to national governments pleading sovereignty, and to corporate interests too. It also draws attention to a particular feature of the situation affecting environmental measures taken by both national and international bodies alike.

That feature is the entire dependence of the actors, or would-be or should-be actors, on scientific information, usually of a highly sophisticated kind, difficult to translate into terms understandable by politicians and administrators or by reporters for the media, indeed by more than a highly select few individuals at large. The information concerned frequently changes its content, is subject to revision without notice and is not always agreed by the scientists themselves. This being the case, who would choose to rely on political organizations whose *raison d'être* is competition with each other, that are led by power-conscious individuals singled out for their capacity to concentrate national feeling, always directed towards diplomatic and military success, and dependent upon, swayed by, media people avid for sensation? This is a particularly needling question at a time when politicians are increasingly dismissed as self-serving by ordinary people and even ineffectual in their attempts to carry out policy. In the USA indeed, and now, it appears, in Australia, there actually exists a fanatical movement directed against all governmental action of any kind. Anarchical perhaps, but certainly not anarchist in the sense that that word is used here.

Framing appropriate policies in relation to scientific discovery, opinion and advice is one more of those formidable, entirely unprecedented problems that the beginnings of the recognition of humanity's place in 'nature' has brought to the fore and which give fascination to our studies.[3] It is not easy to see how effective would be the remedy perpetually urged in such dilemmas as reveal themselves, which is to put the people as a whole in charge, intensifying the democratic process and so eliminating every partial interest, and incidentally rooting out that tendency to corruption so sadly apparent even in the best ordered political systems when the environment arises as a political issue.

A single world government, then, sustained by the familiar democratic apparatus – universal voting, federal provisions, party allegiance and competition – is the picture which seems to come to mind when people reflect on remedying the universal environmental crisis. Highly desirable as this would be it is decidedly utopian and would require a worldwide programme of outright revolution. Moreover, the model itself seems a little conventional. Surely the fresh intellectual horizons which are opening out should persuade us to think again about political forms as well as political assumptions. We must examine every possible mode of conducting collective life, by no means omitting those established among peoples living physically closest to 'nature'. No doubt the more academic and idealistic might wish to revive Plato's doctrine in his *Republic* where intensively educated philosophers would do the diagnosis and the ruling, philosophers with maximal capacity for scientific understanding, along with practical insight and personal responsibility.

It must be noticed here that in so many respects the obsolescence of our political institutions consists not so much in their lessened capacity to do what they have always done as in their now evident inability to cope with the absolutely new, a challenge which will recur in perpetuity. A body of supremely intelligent, perfectly informed, entirely technically competent philosopher-rulers might indeed be what is needed, in perpetuity too, as Plato had argued.

To find ourselves exploring for our purposes something as abstract and impractical, some would also say as tendentious, as the Platonic system, after what has been said here about the necessity of being sensible and hard-headed and close to popular sentiment, may perhaps have to be taken as indicating that the problem in hand is in fact insoluble. Further discussion could lead only to despair, should such be the case.

This is not how we shall conclude. But before the finale appears upon the stage we must make an end of the argument as to the obsolescence of nation-states as well as collections of them like the UN and its agencies. These last institutions along with a miscellany of less official ones, mostly concerned with arbitration, are all that exists in the way of truly global executive and judicial instruments for environmental or any other purpose. All the rest is done by constellations of national governments, some by Big Brother USA acting on its own. So dominated by the nationalist ideology is the sphere of world relationships that no entity other than a sovereign nation-state can plead before the Court of International Justice, the most august of UN agencies, although it appears that commercial corporations can get something of a look-in, especially with the arbitration bodies.[4] World environmental opinion as given voice by organizations such as Greenpeace and Friends of the Earth, therefore, have no direct access to established 'international' justice. Nationalism, crystallized community aggressiveness, dominates all, however modulated it may seem to be.

Once again we find ourselves on the verge of maintaining that nothing effective is possible unless and until everything extant is swept away and a new beginning made by mustering individuals all over the world into a new political association. This is no constructive policy for us as things now are and look likely to be.

There is, however, an emergent feature growing up within the political processes of nation-states which could just conceivably offer an escape route. The opacity of the relationships of self-styled democratic governments with the masses of individuals for whom they are responsible has been vividly illustrated in recent years in the remedial exercises carried out by what are called Citizens' Juries in Britain, and much more rigorously and to our point by the Deliberative Polls undertaken in Britain, the United States, and Australia.

These novel departures have demonstrated that it is indeed open to every single one of us to participate by proxy of a particular kind in instructed, responsible, socially oriented deliberation on national public affairs, and if national, why not 'international', global affairs?

Provided that the myriad linguistic and organizational problems were resolved, there could conceivably exist something like a notional world assembly, so to speak, of participating persons, deliberating not as citizens of nation-states in contact with each other, but as members of the whole order of humans. That assemblage, should it ever come into existence, would be intermittent, lasting no longer on any one occasion than a television programme broadcast and received quite literally world-wide. It would also be virtual, consisting of a random sample of all the world's adult individuals, some hundreds of them in total, assembled in one place with access to all the instruction and information which could be managed, and deliberating together as a face-to-face society. The participants, moreover, could be watched as they talked by any individual with access to a television set, and the number of these is already a high proportion of the population of the world. In the not too distant future it will likely include all humanity who could share the thoughts, adopt or resent the attitudes of the deliberating sample of itself, deliberating on environmental issues whenever occasion arose. This is what happened on a national scale in the deliberative polls held during the US presidential election of 1996 and the British general election of 1998 (see Fishkin, 1991 and 1997). That order of all humans represented by a deliberating world sample would not necessarily be what we define as a secular order, and here the topic of my exposition changes key for a paragraph or two.

Much has recently been made of the growing approximation of speculative thought about the ethics of the environment with religious intellectual systems. In *Environmental Values*, the leading journal in the field, this tendency has come closer and closer to the surface since its foundation early in 1994. Finally in April 1997 *Environmental Values* gave rise to a new periodical *World Views, Environment, Culture, Religion*. The first contribution by Mary Evelyn Tucker was headed 'The Emerging Alliance of Religion and Ecology', appearing along with 'The Vedic Heritage for Environmental Stewardship' and 'The Varieties of Ecological Piety'. In early 1999 *World Views* was in the third number of its second volume, the religious or spiritual themes proliferating.

Now, no intellectual or cultural historian can harbour any doubt of the difficulties in the way of the emergence of a religious-type belief system being adopted by concerned individuals over the whole globe, and expecting just and effective environmental actions from it in any direct way. The precedents are not all encouraging. Religious history has always and everywhere, but perhaps particularly in the West, been marked by fierce ideological prejudices, of racist antipathies, of crusades and conquests, particularly when religion is a constituent of the belligerent ideology of the nation-state. Who knows what would happen if strong religious commitment entirely dominated environmental attitudes? But the religious tendency is present and could become that overwhelming force of belief that its early devotees would like to see: a religion of the environment.

It has to be clearly recognized that if the deliberative arrangement speculated upon here should ever become a practicality, environmental thinking in the religious mode would not require access to the means of mass solidarity and aggressiveness, or to collective power of any kind. It is improbable that such thinking would be revelationist or scriptural to any great degree. It would be unlikely to have much in common with existent religious cults, but no obvious reason to be hostile to them, and would in no way be inconsistent with rational, secular beliefs and attitudes, which would assuredly persist.

Nevertheless, my personal judgement is that the revelationary phase of the environmental faith is already upon us. It began with me as much as 50 years ago on the day in 1949 when, sitting with Fred Hoyle in a BBC studio, I heard him describe space flight as it would be, and indeed was about to be. Our view of who and what we are, where we are and to what we belong, said Fred, will be transformed for ever once we are able to see the earth from space (see Hoyle, 1950).[5] And so it has come about. The sight of that infinitely lonely, tiny pale blue globe hovering in limitless space, with a faint scratch and stain or two on its surface, the solitary evidence of the work of man, but bearing the potential of destruction of itself and its minuscule celestial habitat, has indeed been revelatory. A novel spiritual reality has come to press upon us all.

On the political side of the question the simultaneous growth of intense micro-nationalism within ever larger macro-national, overarching bodies has already been mentioned as significant of the obsolescence of established institutions. Neither extreme gets at all close to providing an efficient and effective instrument for our global, everlasting purposes. But a very small polity like Iceland, with a citizen body of not much more than 100000 begins to resemble a face-to-face society where the interplay of individual opinion really counts. It is an eminent quality of the random sample of persons who come together in a deliberative poll that it does form a genuinely face-to-face society, with a psychology different in order from that which informs the political consciousness of the nation-state.

As we have seen, a deliberative poll drawing its members from the whole population of the world, if ever such an event or continuing series of events could be brought about, would enable the world as a world to deliberate by proxy as a face-to-face society on matters which affected every single human. The national governments and the international organizations, the politicians, the administrators, the party bosses, the media people and the propagandists would see every issue of environmental significance in a new light. In an exceedingly notional way the world itself would be speaking to all of them, and to all of us as well.

This may be an extravagant notion or a fantasy but it could bring each and every person in the world up against any given issue that 'nature' presents. Deliberated opinion and policy choices could be elicited from individuals, as individual citizens of the world, without any intervention whatever by the agencies of the nation-state and with no danger of manipulation by corporations, the media, political personages or

environmental crackpots. As for the execution of environmental policies elicited in this way, that would still have to be carried out by existent authorities, micro and macro, obsolescence being provided for by
all of us doing our duty and making our suggestions as deliberating,
democratic citizens. It is in these directions that the environmental
faith would make its presence felt.

The consideration of obsolescence in our institutions in relation to
'nature' and the environment has brought us a very long way. But it
has not led us to suppose that what has now to be worked for is a revolutionary change towards a new set of political organizations to replace
the obsolescent ones. Nor has it taken us into the philosophical problems raised by the subject, which have been left on one side in this
introductory essay. It will be noticed that the very word 'nature' is in
quotes and the varieties of ecological spiritual experience have gone
without mention – 'non-anthropocentric ecology', 'deep ecology' and
the rest.

Conclusion

We have come to an end by forecasting a spiritual awakening and by
discussing and recommending a technique rather than a programme of
political and intellectual change, though both are implied by the statements which have been made. It is a technique which, if it could be
implemented, might permit us to think our disillusioned thoughts
about extant political organization and its fundamental limitations in
respect of our relationship with our earthly habitat, secure in the
knowledge that the totality of inhabitants of our planet had at last
acquired a voice of its own which the powers-that-be could not ignore.
Some might say that this hoped-for solution is no more realistic than
global revolution, or even the Platonic utopia, notwithstanding it is so
much closer to the interests and outlook of ordinary people.

However this may be, clearly finding any way to take account of the
global opinion that environmental matters require will be an exceedingly complex and lengthy undertaking. If trying to find a way to
adopt the dual scheme sketched out here makes us into anarchists,
anarchists of a peculiarly contemporary kind, so be it.

Notes

1. This would have extended the theory of contract so widely accepted as an
   explanation of social and political authority in an earlier Europe to questions

of intergenerational obligation, called by Laslett the intergenerational tricontract.
2. Scathing as this book is as to current discussions of sustainable development it recognizes and goes far to decide the central issue, securing that justice shall obtain between us as we are now and our successors as inhabitants of the earth.
3. This point should be well appreciated in Britain where in recent years and months the character of scientific opinion and advice has been much to the fore in relation to BSE and genetically modified foods.
4. International law (the only available title of course) is in any case an undeveloped area as compared with other legal systems, with little recent reflection and reform.
5. I was the responsible producer of his famous series of broadcast addresses.

Part III

Humane Government for the Environment

12
Environmental Justice and Global Democracy
Wouter Achterberg

Introduction

The existence of transnational or even global environmental problems raises broadly ethical issues of the first importance for global governance. But in this regard the ethical basis and the form of global governance are also in question. To the ethical basis belongs at least environmental justice: this is in agreement with the generally shared conception of sustainable development. The form of global governance should be democratic, or so I will argue. But what conception of environmental justice could be ethically acceptable to all concerned, rich and poor, living now and later; and is its relationship to (global) democracy more than just contingent?

In this chapter I first argue my way to a broadly liberal-egalitarian conception of environmental justice which should be ethically acceptable to the constituencies mentioned. Next I explore the relationships between (global) democracy and environmental justice. In the last section I discuss critically one model of global democratic governance: Held's conception of cosmopolitan democracy, especially in view of its implications for the realization of global environmental justice (Declaration of Rio de Janeiro, 1992: Agenda 21).

The shape of environmental justice

The leading question here is: how to distribute access to and control over natural resources between people of the present generation, living here or in other parts of the globe, and between us and future generations. That is to say, in environmental justice, intragenerational justice (especially justice between rich and poor countries) should be intimately connected with intergenerational justice. The conception of environmental justice I elaborate in this section is therefore broadly in agreement with the conception of sustainable development internationally accepted, at least since the Earth Summit.

What might be the appropriate starting-point for a conception of environmental justice? Given the diversity of theories of justice in political philosophy, let alone the cross-cultural variety of conceptions of justice, it is important to start from what one hopes will be common ground to all those participating in a rational debate on principles of justice which will regulate the cooperation necessary for a just, sustainable and democratic global order. This may be possible because our focus on access to and control over, particularly, natural resources is or should be a concern common to a broad spectrum of theories or conceptions of justice.

Starting from common ground is, of course, important for a democratic process of planning a world order embodying a measure of environmental justice. In what follows I make use of some basic ideas drawn from philosophers working in a broadly liberal-egalitarian tradition.

The ideal point of departure, I submit, is Dworkin's 'abstract egalitarian thesis' which states that '[f]rom the standpoint of politics, the interests of the members of the community matter, and matter equally' (Dworkin, 1983: 24). He conceives of this principle of abstract or fundamental equality as the 'egalitarian plateau' (1983: 25). All serious modern political theories find themselves on this plateau, at least to begin with, and from it they must derive the political, social and economic conditions under which the members of the community can be treated as equals.

This plateau, from which many different destinations can be reached, might seem too abstract to generate specific conclusions. But the just distribution of natural resources, in a broad sense, also includes functions such as assimilation of waste, the sense in which environmental scientists speak of 'environmental space'. This is the main subject to be treated in terms of the egalitarian plateau and it will be seen that, starting from it, sufficiently specific conclusions about it can be generated. However, how does one proceed from here?

Starting from the egalitarian plateau we can say first of all that everybody should have a rightful share of the material benefits of the natural resources on the planet. This rightful share might come in two different ways: either as an equal part of the material benefits of the natural resources or as a right to enough material benefits to meet at least basic needs. Let us look at some appropriate examples of both ways in turn.

The first way has some affinity to the moral idea that equal respect and concern is owed to members of future generations too. A good example is the view of Luper-Foy (1995). He argues for a principle of inter- and intragenerational equity, called the *'resource-equity principle*: resources are to be handled in a way that is equitable both across the globe and across the generations' (Luper-Foy, 1995: 96). He infers that intergenerational equity 'reduces to the demand for indefinite sustainability in the areas of both pollution and consumption' of natural resources (Luper-Foy, 1995: 96). The rate at which people reproduce makes a difference, though. The rate of reproduction should also be indefinitely sustainable.

The 'resource-equity principle', then, has to be generalized to the 'sustainable consumption-reproduction principle'. The latter principle is that 'each generation may consume natural resources, pollute, and reproduce at given rates only if it could reasonably expect that each successive generation could do likewise' (Luper-Foy, 1995: 98). Intra-generational distributive justice is constrained by intergenerational justice in the following way: by 'setting the ceiling on the resources that are available to each generation, and by delineating *how* resources may be consumed, the sustainable consumption-reproduction principle specifies what resources are available to us as a generation' (1995: 100). I will not discuss here Luper-Foy's very radical institutional proposals to realize his principles.

The second way is an expression of environmental concern to a lesser extent than the first. Its basic idea is that an equal part of (the value of) the natural resources may or may not suffice to meet basic needs. But that is precisely what this second way wants to guarantee: a share of material benefits that is enough to satisfy basic needs, that is a standard of living at least adequate for the health and well-being of oneself and one's family including food, clothing, housing and medical care. This way, focused on the position of the worst-off, leads to Pogge's (1995) proposal of a Global Resource Dividend (GRD).

The GRD is not intended by Pogge as a complete criterion of global justice, but as an institutional proposal taking a first step in the direction of realizing an egalitarian conception of global justice. The background of the GRD is the awareness of crushing poverty of about one-fifth of the world population. There are also very many people affluent enough to help in relieving this poverty substantially.

Pogge argues that the resulting inequality is radical and of social origin. It is, moreover, unjust, and we, citizens of affluent countries, have in this regard a negative responsibility on three counts. In the first place, because we participate with the poor in one global system of social institutions that generates poverty, avoidable poverty at that. Next, because we, the better off, draw substantial material benefits from the use of natural resources, while the global worst-off are to a considerable extent and without compensation excluded from these benefits. Third, because the radical inequality between rich and poor is the result of a historical process fraught with massive crime (colonialism, slavery, and genocide).

A GRD, funded by a tax on the use of natural resources and to be paid by the governments of the using (harvesting or dumping) countries, of about one per cent of global social product, yields the sum of US$300 billion per year. This is enough, as Pogge hopes, to eradicate 'global poverty within one or two decades' (1995: 183). The GRD, then, is funded by a tax on consumption and not just on the owning of resources, but ultimately the tax-base is environmental consumption. This is because the 'tax falls on goods and services roughly in proportion to their resource content: in proportion to how much value each takes from our planet' (Pogge, 1994: 200). All use of the environment is taxed, not only use by the rich. The proceeds of the tax should be used to advance the emancipation of the worst-off in our world, so that at last all will have a standard of living adequate to meet their basic needs in dignity.

This 'modest' proposal (Pogge, 1995: 192) towards institutional reform of the global order takes the existing order of sovereign states for granted and leaves them the control over the resources on their territory. The term dividend, however, is meant to express the belief that all humans have an inalienable right to a share in the material benefits of the use of natural resources.

The implementation of the GRD scheme doesn't require a world government or even a centralized agency applying sanctions. A decentralized regime, supported by the major economic powers, might suffice to enforce the obligations under the GRD scheme.

The next step is to compare the two ways distinguished above in order to find a reasonable order of priority or a balance between them. The GRD regime is designed to abate global poverty in the first place and therefore gives pride of place to intragenerational justice. Intergenerational justice is subordinate to this aim but its requirements can be met to a certain extent by appropriate adjustments of the GRD. Thus Pogge deems it necessary, in specifying the shape of the GRD, to take the interest of future generations in a rich and healthy environment into account. This could be done, for example, by taxing the use of natural resources that may soon be depleted or a use of environmental functions that leads to long-term damage to the environment (Pogge, 1995: 196).

Luper-Foy tends to consider intergenerational justice as a precondition for international justice, but modifies his resource-equity principle to take the effects of population growth into account, which leads to his 'sustainable consumption-reproduction' principle.

Reviewing both proposals, I am inclined to say that from the point of view of intergenerational justice, global resource use should stay within the boundaries of ecospace and that the tax-base and the level of the GRD should be adapted accordingly. This, at least, is what I would propose if intergenerational justice is to be taken seriously. As the Commission on Global Governance puts it:

> Equity needs to be respected as well in the relationship between the present and future generations. The principle of intergenerational equity underlies the strategy of sustainable development, which aims to ensure that economic progress does not prejudice the chances of future generations by depleting the natural capital stock that sustains human life on the planet. Equity requires that this strategy is followed by all societies, both rich and poor.
>
> (1995: 52)

But how seriously, with what moral weight, should sustainable development in this sense be taken? 'Taken seriously' here cannot mean simply overriding in all circumstances. So, how should we balance the claims of intragenerational justice (as specified by Pogge) and those of intergenerational justice (as envisaged by Luper-Foy and the Commission)?

The effort to realize intragenerational or international equity within the limits of sustainability (in the sense explained) might make the proceeds of the GRD lower than they could be without taking this precondition into account. This is because the rate of exploitation of the environment on which the revenue depends would be reduced. Lower proceeds of the GRD would, of course, be undesirable from the point of view of eradicating poverty 'within one or two decades'. Given the urgency of eliminating global poverty in the foreseeable future, the lowering of the GRD is justified, I submit, only if this is compensated for elsewhere. This is where the next part of the conception of environmental justice comes in.

For the nature of the compensatory guideline, I draw upon Henry Shue's (1992) proposal. Although it has been developed in view of the question of what justice would require in climate negotiations between rich and poor countries, I believe his results – especially his distinction between preventing and coping with harmful global environmental change – can be generalized.

The basic idea is that, on the one hand, it is morally right that even poor countries pay for their actual use of environmental space, and so contribute fairly to any scheme to maintain or protect the relevant functions of environmental space.

On the other hand – and this concerns the case of coping – it would not be fair to ask especially poor countries to pay additionally. For example, it would be unfair to ask them to slow down their qualitative or quantitative consumption of environmental space in order to solve ecological problems that can no longer be prevented (e.g. climate change due to lavish use of fossil fuels), and which is predominantly due to the past activities of rich industrialized countries. In fact, it would even be doubly unfair (Shue, 1992: 91) because of an unjust international order (background injustice) and because the rich nations have caused most of what makes problems such as global warming and ozone depletion so troublesome.

For the very poor countries coping with inevitable environmental harm using their own resources is life-threatening. Why is this?

> Because in very poor nations almost all big problems are life-threatening. This is what it means to be very poor: it means having no cushion to fall back upon, no rainy-day fund, no safety-net, no margin for error. Being very poor means living on the edge, and having a big problem – sometimes, even, having a small problem – means going over the edge: losing one or two of the children, for example.
> (Shue, 1992: 393)

In short, their 'vital interests – their survival interests' are at stake (1992: 394). What follows from all this?

As a first and very minimal step – of justice – that the poor nations' costs of coping be borne by the rich countries. The financial aid and transfer of technology by them 'ought to be sufficiently timely and substantial' that the poor countries don't have to

> sacrifice in any way the pace or extent of their own economic development in order to help prevent the climate changes set in motion

by the process of industrialization that has enriched others ... Even in an emergency one pawns the jewellery before selling the blankets. The weak guideline being proposed as a start merely reflects that, whatever justice may positively require, it does not permit that poor nations be told to sell *their* blankets in order that rich nations may keep *their* jewellery.

(Shue, 1992: 394–7)

In sum, the following guideline suggests itself: if one accepts that the tax-base of the GRD ought to remain within the boundaries of environmental space, then only under the proviso – let us call it: 'Shue's proviso' – that one also accepts that the poor countries don't sustain opportunity costs in terms of the pace or extent of their own economic development, compared with an environmentally less constrained GRD. My proposal for a conception of environmental justice which balances intergenerational justice, conceived in terms of sustainability, and intragenerational justice, particularly as aimed at abating poverty, is therefore to take sustainability or, more precisely, the sustainable consumption-reproduction principle as a precondition for abating poverty by means of a GRD, but only within the constraints of Shue's proviso.

From environmental justice to democracy

The connection between environmental justice and democracy is not at all necessary or even clear. Democratic deliberation or procedures need not lead to a consensus about environmental justice, of whatever variety. And, theoretically at least, environmental justice might be embraced and implemented by a benevolent dictator, though this, perhaps, is far-fetched and, anyway, the environmental record of known authoritarian regimes has been very miserable. However, it is possible to indicate some connections between environmental justice and democracy.

A viable conception of environmental justice presupposes a broad, democratically reached agreement between those affected. And, on the other hand, a viable democracy presupposes an environmentally just global order. How plausible are these supposed relationships?

Starting with the first of these assertions, note that the conception of environmental justice developed so far is itself not at all the obvious or the only possible result of the balancing act performed earlier. At several points reasoned decisions had to be made to reach an unambiguous conception and it was not clear beforehand in what direction the balance of reasons would point. What is more, a number of equally legitimate and plausible conceptions of environmental justice might have emerged from deliberation. Given the interests at stake and the conflicts of interests and values involved, only democratic decisions, reached after due deliberation and based on rational consensus between all affected by the implementation of the policy guided by the conception, may legitimate a conception of environmental justice as a guideline for environmental policy.

A second problem is that the limits within which environmental consumption should stay to be sustainable, i.e. the boundaries of environmental space, are themselves not based on any 'hard' or objective constraints. It is misguided to expect a purely scientific specification of this space. As two Dutch environmental scientists put it,

> [the] most fundamental reason is that making pre-scientific choices and introducing value judgments are inevitable regarding several issues such as: which biospheric elements to preserve and at what levels, which degrees of risk can be taken, and how to handle uncertainties and lack of knowledge.
>
> (Weterings and Opschoor, 1994: 227)

Obviously this doesn't mean that anything goes in the debate on the scientific, normative and policy implications of sustainability. Some arguments are still better and some reasons stronger than others, as I hope this chapter will make clear.

The impossibility of specifying purely scientifically or objectively what the goal of sustainability implies gives us a fundamental reason, and not just one from political expediency, why the process of planning for a sustainable – that is, environmentally just – society should be a democratic one. In other words, why this process should not be top-down and expert-dominated, but to an important extent bottom-up and based on broad participation in the processes of deliberation and social learning to which, ultimately, the collective effort to establish sustainable patterns of production and consumption worldwide comes down.

A big problem here is that the very concept of democracy is also essentially contested. Although I cannot argue for it here, the conception of democracy that would seem most appropriate to the normative decisions at hand is presumably (one or another variety of) deliberative democracy.

A viable and socially legitimate conception of environmental justice seems to imply a strong and vibrant (deliberative) democracy. What about the reverse implication? Does democracy also imply the realization of environmental justice (as elaborated on p. 183)? At least a viable global democracy seems to. As the Commission on Global Governance puts it:

> Societies in which there are deep and expanding social or economic disparities face enormous obstacles, whether in creating or maintaining democracy. Citizens who must struggle daily to meet basic needs and who see no possibility of improving their circumstances are unlikely to have either the interest, or the ability, to work on behalf of democratization. To be sustainable, democracy must include the continuing prospect of contributing to the prosperity and well-being of citizens.
>
> (1992: 61)

In general, of course, one cannot reasonably expect a stable democracy in regions where many people are most of the time struggling to meet their basic needs and anyway lack the minimal information and education necessary to participate in the democratic process. But the implication also has aspects of prudence and morality, as Pogge (1994, 1995) points out.

The affluent parts of the world will increasingly be exposed to threats and dangers from the poorer and unstable regions. These negative externalities include, according to Pogge, attacks by armies or terrorists, illegal immigration, drugs traffic, epidemics, pollution and climate change, fluctuating prices on world markets and scientific-technological or cultural innovations. But radical inequality and the resulting instabilities endanger the security of posterity and the survival of society and democracy, not only in the affluent countries, but worldwide. So we have, concludes Pogge, also a moral interest in peace, i.e. in a world in which a diversity of societies and cultures enjoys a long-term prospect of peaceful coexistence and mutual enrichment.

Democracy as discussed so far is a rather underdetermined affair. Public deliberation should be important in it, but what more can we say? It has to be a kind of *global* democracy because a legitimate and effective realization of environmental justice requires global cooperation and also because the processes of economic, environmental and cultural globalization don't leave us much choice. Should the global order establishing the sustainable GRD scheme and implementing

Shue's proviso be a liberal democratic order? To judge from the environmental record – up to now not very impressive – of existing liberal democratic nation-states, I believe the answer has to be 'better not'.

The structural problems of liberal democracy, the long-term consequences of environmental degradation, are well known. I am not going to rehearse these here but will instead ask what a more radical model of global democracy, which enhances and broadens liberal democracy, might be like, and what its promise for the realization of environmental justice may be. The most detailed model of global democracy so far is the conception of cosmopolitan democracy developed by David Held, which I shall discuss critically in the next section.

Cosmopolitan democracy

Held's (1995, 1996) model of cosmopolitan democracy is, of course, prescriptive, not descriptive; moreover, it is a radical proposal. But it does not exclude other more modest proposals such as Pogge's. As I conceive models of global democracy, they need not exclude each other. That is, the institutions and procedures embodying the normative principles that underlie the models need not come in all-or-nothing packages. And that is just as well, given the urgency of the problems, political feasibility and the desirability of international fora of public deliberation. The possibility of correcting one or more of the well-known structural problems of traditional liberal democracy with environmental degradation also calls for attention in constructing a global democratic order.

Held's proposal for a cosmopolitan democracy is less detailed than other proposals but his theory also tries to transcend the existing international system in a more radical way. I will first summarize some main points of Held's theory and next trace the consequences for environmental justice.

The movement towards cosmopolitan democracy is a process that deepens and extends democracy across and between nations, regions and global networks, thereby entrenching 'democratic autonomy on a cosmopolitan basis' (Held, 1996: 353). The normative linchpin of the theory is the principle of democratic autonomy:

> people should enjoy equal rights and, accordingly, equal obligations in the specification of the political framework which generates and limits the opportunities available to them; that is, they should be free and equal in the determination of the conditions of their own

lives, so long as they do not deploy this framework to negate the rights of others.

(Held, 1995: 147)

The question now is: how can the basic ideas thus expressed, self-determination of the people and limited government, also be applied transnationally, in a global democratic order? Nation-states will not be abolished by cosmopolitan democracy, but their sovereignty will be limited by a new layer of democratic governance at the regional and global level. Citizenship will accordingly be extended.

Thus, a multilevelled transnational democratic community, or a 'transnational, common structure of political action', will take shape, but under the guidance of cosmopolitan democratic law, necessary to the achievement of autonomy. This law is to be embodied, in the long term, in a 'new Charter of Rights and Obligations locked into different domains of political, social and economic power' (Held, 1995: 279).

The transformation of the existing (United Nations) system of global governance in order to entrench cosmopolitan democracy has both long-term and short-term aspects. The short-term objectives such as reform of the UN Security Council and creation of a UN second cham-ber (directly elected) are modest. This is less so with the long-term objectives, which include, among others, a global parliament, a global legal system and an international criminal court, and the establish-ment of the 'accountability of international and transnational eco-nomic agencies to parliaments and assemblies at the regional and global level' (Held, 1995: 279).

But Held also extends the democratization under cosmopolitan law to the sphere of civil society and the economy, thereby transcending the framework of liberal democracy. For example, he proposes system-atic experimentation with economic democracy (short term), 'different forms of the ownership and control of capital' (1995: 266), and a social framework to determine investment priorities on the basis of public deliberation and government decisions (both long term). On the other hand there remains an 'extensive market regulation of goods and labour' (1995: 280).

What implications does Held's model have for environmental protec-tion and (sustainable) development? He doesn't say much about these problems but he does include a right to a 'clean, nontoxic, sustainable environment' in the category of 'health' rights in his system of rights, which are 'necessary to enable people to participate on free and equal terms in the regulation of their own associations' (1995: 191–2). He says nothing about intergenerational justice, nothing about the relations between the affluent and the poor in the global system. But perhaps we can infer what more he would or could say if he had given thought to the matter.

Held categorizes rights in seven clusters, concerning: the body, welfare, culture, civic associations, the economy, coercive relations and organized violence, and legal and regulatory institutions. These are general human rights underwritten by cosmopolitan law. From this, one can infer that the poor on this planet are at least rich in legitimate grounds to claim in the new second chamber of the UN or before the Human Rights Court that they be enabled to enjoy those rights. Development aid should be forthcoming, then something like the GRD should be established.

Why the GRD? Because it is much more in agreement with the spirit of the principle of autonomy than traditional development aid which often keeps people dependent. Let us look next at one of those rights, the right to a 'clean, nontoxic, sustainable environment'. This, as a general human right, is also a right that belongs to future generations of humans. Therefore, we, living now, have a duty to do something so that people living later can enjoy a clean, nontoxic and sustainable environment. So the cosmopolitan law justifies pursuing sustainable development.

It is difficult, though, to derive from it any clear indication concerning the relative moral weight of inter- and intragenerational justice. However, Held acknowledges in his seventh category of rights a right to 'adequate and equal opportunities for deliberation' and he also proposes as a long-term 'cosmopolitan' objective the 'public funding of deliberative assemblies and electoral processes' (1995: 279). We can therefore be sure that, after ample deliberation, a just and democratic solution to this remaining problem will be found. If we add Held's inclination to democratize to a certain extent the economy, therein transcending liberal democracy, his model of global democracy seems to satisfy the requirements of realizing environmental justice. What more could one wish for? At least one would like to know more about how to go from here to there.

Even to realize the short-term cosmopolitan objectives mentioned earlier seems a tall order in the present international system. Held calls cosmopolitan law the 'framework for utopia' (1995: 266). He calls his own theory, or likens it to, 'embedded utopianism', which 'must begin both from where we are – the existing pattern of political relations and processes – and from an analysis of what might be: desirable political forms and principles' (1995: 286).

But is the transition from one to the other – itself necessarily a democratic transition – feasible given the political, economic, social and environmental constraints of the present world order? What possibilities of democratization does this world order offer? And what are the costs of making the transition one way or another in view of the urgency of the problems of poverty and global environmental degradation?

Modest proposals like Pogge's take the constraints and possibilities of the present international order more seriously than Held's while their long-term objectives are presumably no less utopian than Held's. The first steps, then, on the way to global governance which is democratic and environmentally just should be like those proposed by Pogge and others.

Conclusion

This chapter has elaborated a conception of environmental justice which in a sense gives pride of place to the abatement of world poverty but which nevertheless should be ethically acceptable to all concerned and should therefore belong to the ethical basis of global governance. Moreover, the argument has shown that the realization of environmental justice in this sense implies global democracy in one form or another, and that global democracy presupposes an environmentally just global order.

One radical democratic model of global governance has been discussed: its ethical basis turned out to be acceptable and to have affinity in its starting-point with the conception of environmental justice argued for. Nevertheless the model was found underdetermined in its implications for the sustainability of a democratic global order and its indications for how to get there in an environmentally benign way.

13

The Politics of Cosmopolitical Democracy

Daniele Archibugi

Introduction: the state as the centre of political authority

At the dawn of the twenty-first century, if we stop to ask ourselves which political institutions are the depositories of power, we are forced to give the same answer a seasoned observer would have given in 1815: namely, states. In the course of the last two centuries, in fact, states have asserted themselves increasingly as veritable oligarchies of world politics. You only have to look at a political map of the world to grasp the fact. With the exception of Antarctica, the entire surface land mass of the planet is delimited by states. To make this pre-eminence all the more evident, geographers use bright colours to show who the owners of the world's surface are. Inside states, the colours are homogeneous. If the surface of the United States is green, that of Canada is red. It may be a formality, but it does testify to the pre-eminence of states over individuals. The attributes of individuals become secondary inside states. Whether their skin is white or black, whether they originate from Europe or America, whether they are Christian or Muslim, they are considered, first and foremost, from the political point of view, Americans or Canadians.

It is states that have armed forces, that control police, that mint currency, that allow individuals to cross their borders, that recognize citizens' rights and impose their duties. To assert their dominion over individuals, states have had to use a variety of more or less coercive means, such as armies, the police and the public administration. In one chunk of the world, which has fortunately grown larger and larger, state political communities have arisen in which the use of power is tempered by so-called checks and balances to limit the abuses which a concentration of coercive means might lead to. Thus, since the formation of states, a slow, complex interaction began between those who held power and those who were subject to it. The state evolved under the pressure of citizens to become not only a tool of dominion, but also a service structure.

Never in the history of the human race has there been such a successful structure, one which has, *de facto*, become of crucial importance for all the inhabitants of the planet. No single religion or even all religions put together have ever centred as much power on individuals as states as a whole possess today.

Since they were born, states have had to come to terms with the heterogeneity of the individuals inside them. Individuals speak different languages, have distinct traditions, profess different religions and belong to different races. In a word, they belong to groups with different identities. Some states are more homogeneous than others, but not one of them can consider itself to be totally homogeneous.

In the course of the centuries, every state has used a variety of means to pursue a greater degree of homogeneity. Some states have sought to found their own national identity on religion, others on language, others still on blood or race. The nation concept, which is not to be found in nature, has served precisely for this purpose. Through wars, revolutions, treaties and negotiations, states have changed their borders, provoked exoduses or promoted fusions, often to render their populations more uniform. Attempts have been made to convert populations to the dominant religion or to root out vernacular languages and, if this proved impossible, the 'die-hards' have been expelled or even repressed.

Other states still, the most enlightened of all, have looked for institutional devices to regulate diversity. Such states have opted for equality among different religions and, as early as two centuries ago, this kind of norm was being endorsed in the first constitutional charters. On more than one occasion, attempts have been made to stir up a spirit of nation and homeland by fomenting sentiments of nationalism and patriotism. A number of tools have been widely used to achieve this end: the foreign menace, internal danger, the creation of a cultural identity founded on the values of the flag, culture and the arts, sports teams and television have in their own way helped to make states more cohesive.

Despite assembling this daunting collection of tools, the state has nonetheless failed fully to assert its sovereignty, either internally or externally. The external sovereignty has been constrained by international power politics. A few states only have been allowed to be independent and could avoid accounting for their choices to other, and more powerful, states. On a few occasions, state sovereignty has been violated by open military interventions. Much more systematically satellite states have had to deal with the interference and requirements of their stronger neighbours.

Since its origins, state sovereignty has also had to fight against a dangerous hidden internal adversary: namely, the fact that neither civil society nor nature respects the borders which it has artificially created. Societies have thus entered increasingly into contact with one another, and this has inevitably made their borders permeable. Men and women love travelling and describing what they see, imitating what their neighbours do and allowing themselves to be convinced and even converted. Economic society is founded on the exchange of different goods and the ones that are scarce in a given place become more precious there. Not that states have systematically opposed the permeability of borders. Only the most obtuse, dominated by despotic regimes, have attempted to prevent their subjects from travelling abroad and seeing what life is like elsewhere. In the majority of cases, state institutions have facilitated international exchanges and even set up bodies to make them possible (see the vivid account of Rosenau, 1997). For example, to cross borders it is necessary to have a passport, to trade it is necessary to have the authorization of the customs authorities, to transfer capital it is necessary to have the permission of the currency authorities. Until a short time ago, authorizations were needed even to translate books and profess religious beliefs different from the established ones. The apparatus of norms and permits that the state endowed itself with was basically a sign of its attitude towards the individual: you are mine, the state authority seemed to warn, but I benevolently allow you to satisfy your curiosities and to see what happens outside my borders.

To reduce fractions resulting from the artificial subdivision of the land into territorial states, transnational exchanges have been 'oiled' by creating a variety of bilateral agreements and setting up multilateral institutions. The great success of the institution of the state can also be explained by the fact that it has been versatile enough to absorb and regulate even what happens outside its borders. Sophisticated juridical constructions, such as those of international law, the existence of diplomatic structures and the birth and development of intergovernmental organizations are just part of the impressive collection of tools that states can pull out to regulate relations among themselves.

The state and globalization

For many years now, however, the political organization of land founded on states is beginning to show signs of yielding. Not that cracks have begun to appear all of a sudden; far from it. There is no reason to believe that the state system will collapse as the Roman empire did, and probably many critics of the state exaggerate the size of the cracks that have opened up. However, irrespective of the depth of the present crisis, it is evident that many of the problems of the political organization of contemporary society go beyond the scope of single states.

In the first place, a significant number of the problems which states have to address is outside their autonomous jurisdiction. The planet is experiencing a phase of strong and ever growing interdependence: the US Federal Reserve's decision to raise the interest rate may provoke a substantial rise in unemployment in Mexico; the explosion of a nuclear power station in the Ukraine triggers environmental disasters throughout Europe; the lack of prompt information about the diffusion of AIDS in Nigeria may cause epidemics in many countries in the world (the impact of globalization on national political community is emphasized by Held, 1995: 99–136). In all these circumstances, sovereignty inside states is not called into question by armies, missiles and armoured cars, but by elements which spontaneously escape the control of national governments. This is the process which, for some decades now, we have been calling globalization. This is not the place to discuss just how significant globalization is, or the extent to which it is a recent phenomenon (the debate is addressed in a growing literature: see, for example, Held, McGrew, Goldblatt and Perraton, 1999). It seems natural, however, for states to remedy the situation, though the traditional response, that of creating special intergovernmental institutions with a mandate to manage and mediate specific international systems (such as trade, industrial property, nuclear energy or epidemics) is only partly capable of serving the needs of society.

In the second place, in the course of the 1980s and 1990s we have also seen a critique of the state from the inside. I no longer refer to the classic revolutions of the modern era, whose fundamental aim was to replace one government (or form of government) with another. True, the French and Russian revolutions questioned many things, but never the very existence of the states of France and Russia. Today, instead, we see growing dissatisfaction among peoples who believe that their political community is too centralized for their needs. We have thus seen some peoples claim the constitution of states on a smaller scale (this is the case of the myriad of states that have sprung up since the dissolution of the Soviet Union and the Yugoslav republic and the separation of the Czech and Slovak republics). In other cases, political groups keen to achieve greater local autonomy or even to secede from their state of origin have gained renewed strength. In Canada, Spain, Great Britain, Italy, and Indonesia we have seen separatist groups come into being and consolidate their role. Then we have the painful phenomenon of peoples who still claim their constitution as a state and feel oppressed by the state they belong to.

The interstate system has so far struggled to provide adequate political community for Kurds (so far without success), Palestinians (with only very limited success), the inhabitants of East Timor (belatedly and at enormous cost) and for many other peoples. In the future, these problems are likely to become less important than that which has only just appeared with globalization: communities of immigrants. Migrations have made new settlements in traditional states increasingly important. In modern cities whole communities with a language and culture of their own have taken root – Turks in Berlin, Chinese in Los Angeles, Arabs in Paris, Indians in London, Vietnamese in Montreal – posing new problems for consolidated political communities. These are still minorities which do not claim constitution as states, but they do want their own cultural identity to be respected and protected (Kymlicka, 1995: 121–3; Tully, 1995: 183–7). These new cultural identities in the bosom of existing political communities will continue to grow in importance in the course of the next century. Will the state system be capable of meeting their claims?

If we combine the problems posed to the state from the outside by the process of globalization with the internal problems caused by the demand for greater autonomies, the aphorism that the state is too big for small things and too small for big things takes on new value. It is here that emerges a tendency towards a form of world governance stronger than the existing one. But what form must world governance, so often evoked after the fall of the Berlin Wall, take?

Internal democracy and international system

The state has managed to meet the needs of the individuals who form it wherever it has linked them to the management of the *res publica*. The existence of a recognized institution that is the only one authorized to use force legitimately is the precondition for the birth of democracy.

The fact that democracy has extended quantitatively is certainly a great success of the state. And despite all the uncertainties and ambiguities in neophyte countries and the persisting contradictions in the countries of the old guard, democracy is increasingly emerging as a legitimate form of government. It was the people of the world and not academic pulpits which achieved this target. The last decade of the twentieth century will also be remembered for the interminable queues of men and women we have seen in the East and South of the world, waiting outside polling stations in countries where the sacred rite of democracy – free elections – had previously been prohibited.

So to what extent has the new wave of democratization also contaminated the international system? We have witnessed different, more or less stringent forms of regulation of world life. The international system has been managed by threats, wars, accords and diplomacy. International political choices have never been dictated by anarchy alone. But no such form of regulation has ever been inspired by the principles and values of democracy. Transparency of action is replaced by summits between obscure powers, the function of representatives elected by the people by cunning diplomats, and sometimes even by secret agents, and judicial power by intimidation or even reprisal. In the final analysis, it is force – be it political, economic or, ultimately, military – which regulates conflict.

International institutions, such as the League of Nations first, and the United Nations today, have been founded on some of the principles of democracy such as the existence of constitutional charters, the transparency of actions and the institution of an independent judicial authority, but they have been hamstrung by restrictions which prevent them from performing the noble function which their statutes envisaged for them. Democracy has achieved important targets inside states, but very unimportant ones in the international sphere.

To what is this paradoxical contradiction due? How can a system of government have developed so much inside states and so little in relations between states and on global issues? Some people believe that it is impossible to be democratic with others who are not democratic, and that the opportunistic conduct of democracies in foreign policy is caused by the existence of autocratic regimes. This thesis has justified many of the policies of liberal democracies in the dark era of the cold war: troops were sent to Vietnam to check the advancement of Soviet communism, apartheid in South Africa was justified as a means of keeping the 'red menace' out of the continent, and an elected left-wing government of Chile was overthrown to avoid a second Cuba in Latin America. We might have expected a radical change in the foreign policy of liberal states after the fall of the Berlin Wall, but, quite frankly, the signs of such a change are conspicuous by their absence.

The same school of thought argues that if all the states in the world were democratic the problems of war would be solved, the principles of self-determination of peoples and human rights would be respected. Simply adjusting national systems could solve the problem of global democracy. In support of this thesis, it is argued on the basis of new statistical sources that democracies do not fight each other. If all states were democratic, it is argued, there would no longer be wars between states (see Brown, Lynn-Jones and Miller, 1996; Russett, 1993).

A more thorough historical and logical analysis of this thesis shows not only that it is wrong, but also that it is downright dangerous. First of all, it is unclear which countries deserve the licence of 'democratic' and who is authorized to issue the licence in the first place. If this 'licence' is issued by other states, it is evident that the criteria for issuing it will be distorted to favour prevailing interests. To cite a few glaring examples of governments that enjoy the sympathy and antipathy of self-proclaimed 'enlightened states', are we convinced that Indonesia is more democratic than Iraq, Guatemala more democratic than Cuba, or Turkey more democratic than Serbia? And if, following the cues of scholars who have tried to measure the level of democracy in single countries, it emerged that in all those countries democracy was non-existent or only formal, how could we justify the difference in attitude of democratic countries? For example, how come Turkey is a full member of NATO, the military community of Western democracies, whereas Serbia was bombed by it? In the second place, even if all the countries in the world opted for the democratic way, in such an anthropologically variegated planet as ours, some states would always be more democratic than others. The long march towards democracy will always be made by countries that walk at different speeds, and if it really does intend to help them on their way, the international institutional system has to accept diversity.

Finally, there is no proof, either historical or theoretical, that democratic states are more respectful than others of international legality in foreign policy. The United States, Great Britain and France, to cite just three of the industrial powers that boast well-rooted liberal-democratic traditions, do not hide the fact that they defend their own interests in the international sphere. Foreign interventions conducted by democratic states are not necessarily inspired by the principles of their own constitutions: the non-democratic peoples of Indo-China had to struggle for their independence by fighting first against the troops of the democratic French government, then against those of the liberal-democratic American government. The history of democracies is sadly constellated with aggressions against political communities that, albeit not informed by the values of democracy, had the sacrosanct right to their own independence.

As the abuses of colonialism show, democracies have too often been biased in their judgements, even *vis-à-vis* human rights. Great Britain, the United States and France, the three countries of the great declarations of human rights, have respected given human rights with increasing rigour internally, but they have not given a second thought to trampling over the selfsame principles in India, with Indigenous Americans, or in Algeria. Democracy at home does not guarantee that a country will act on democratic principles internationally.

In short, something more than internal democracy is called for. That something may be summed up as the democratization of the international community seen as a process of joining together political communities with different traditions and states of development. This is what a group of scholars have defined as the cosmopolitical democracy project.

Cosmopolitical democracy

The cosmopolitical democracy project is based on the assumption that substantive objectives such as the control of the use of force in international relations, respect for human rights and the self-determination of peoples may be obtained only through the extension and development of democracy (see Archibugi and Held, 1995; Held, 1995; Falk, 1995; Archibugi and Köhler, 1998; Linklater, 1998; Archibugi, Held and Köhler, 1998; Holden, 2000). Contrary to previous work, I have been convinced that the term 'cosmopolitical' should be preferred to 'cosmopolitan'. The cosmopolitical democracy project is therefore more specific than the general approach to cosmopolitanism in the sense that not only does it call solely for a global responsibility, but it is an attempt to apply some principles of democracy internationally. Daily problems such as the protection of the environment, the regulation of migration (see Chapter 5 of this volume) and the use of natural resources (see Chapter 4) must be subjected to democratic control. But for that to be possible, it is necessary for democracy to cross the borders of single states and assert itself on a global level.

Many projects have been put forward in the course of time to achieve a universal republic or world government founded on consensus and legality (for a review, see Heater, 1996). However, it is not easy, either conceptually or, above all, politically, to develop to a meta-state dimension the democratic model that has so far been born and grown at the state level.

To extend the principles and norms of democracy to world level, it is not enough simply to apply on a vaster scale what has already happened in the course of the last two centuries inside single states. Some of the fundamental aspects – such as the majority principle, the unit of norms and the use of coercive power – on which democracy is founded in delimited political communities have to be reformulated, if they are to be applied in global society. This is why the cosmopolitical democracy project stands out from the federalist tradition. The former does not believe that existing states have to be dissolved to give life to a sort of world state. Today, states have a political and administrative function to perform from which there is, realistically speaking, no getting away. Nor can making them bigger solve the present problems of states. On the one hand, extending democracy globally means designing a form of organization of the political community which, unlike the traditional one, seeks to do more than merely reproduce the state model on a planetary scale. On the other, it involves reviewing the functions and powers of states, and depriving them of the oligarchic power that they now enjoy in the international sphere.

What marks out the cosmopolitical sphere is its attempt to create institutions which make the voices of individuals heard in global affairs, irrespective of the voice they have at home as citizens (in democratic states) or subjects (in autocratic ones). Democracy as a form of global governance has to be achieved on three different, interconnected levels: (1) democracy inside states, (2) democracy in relations with other states, and (3) democracy on a global level. As far as democracy inside states is concerned, the problem is to swell the wave that has been sweeping the planet for more than a decade, especially in the half of the world's countries which are still governed autocratically. It is also necessary, however, to beware of democratic fundamentalisms: paraphrasing Robespierre, we have to avoid making peoples democratic against their will. There is a widespread assumption held by some supporters of democracy (or, more generally, by Western politicians) which may be summed up as follows: 'I, citizen of a democratic state, teach you by fair means or foul what you have to do.' Not only is this attitude unbearably paternalistic and ineffective in practice, it is also the very negation of democracy itself. Democracy in fact presupposes the existence of a dialogue among people and among cultures on the basis of equal dignity. The community of democratic states may make an important contribution to the development of democracy in autocratic countries, but that contribution will be all the more effective if it anchors itself to civil society and existing claims and complies with existing rules in international relations.

To promote democracy in interstate relations, it is necessary to strengthen the present network of intergovernmental bodies – the United Nations and its various agencies first and foremost. Numerous proposals have been put forward for reform of the United Nations to increase its democracy (the reform of the General Assembly, of the Security Council, the Court of International Justice, etc.), but too often they have been stonewalled precisely by Western democracies.[1] This shows how little the West is prepared to accept democratic procedures if they look like constraining its vital interests.

Many problems cannot be addressed effectively by intergovernmental organizations alone. On issues such as environmental protection and the defence of fundamental human rights, national state governments do not possess the necessary representativeness for the simple reason that they stand for a community other than the one which suffers the direct and indirect consequences of its choices. It is consistent for the French government to carry out nuclear experiments in the Pacific Ocean, if the advantages all go to France and the radioactive waste harms the peoples of the other hemisphere. There is no 'national interest' of Italy, France or Great Britain, if in Iraq or Iran or Turkey genocide is committed against the Kurdish population; and even if the former states decide to intervene outside their borders, it will be impossible to detect if their actions are due to self-interest or to ethical responsibility. These are the areas in which the need emerges to develop democracy on a global level too, involving the world's citizens institutionally in parallel with and irrespective of their quality as subjects of a given state.

How is it that democratic action moves so slowly outside states? If we observe how well transnational economic interests (suffice it to think of multinational enterprises) and military power (today there is no more efficient institution than NATO) are organized, it is surprising that political parties are still an almost exclusively national phenomenon (Beck, 1999). The Socialist and Christian Democrat Internationals are institutions devoid of effective power, while the Communist International, founded on the idea that the proletariats of the world had common interests to defend across borders, ceased to have a role long before Stalin suppressed it. In Europe, where a single market exists, as well as a parliament elected by universal suffrage, and now also a single currency, parties operate essentially on a national basis. This is the most evident demonstration that forms of political representation have remained locked inside state borders even in this era when civil and economic society operates massively on a transnational basis. This is the true deficit of democracy: the existence of organized interests that fail to correspond to any mandate from citizens (Held, 1995: 16–17).

Today new social and political subjects are appearing in international life. I do not wish to overestimate their importance, but associations such as movements for peace, human rights and environmental protection are playing a growing role in the political process. For the political dimension to exist for the world's citizens, it is necessary for appropriate institutional channels to open up. This is the objective that marks out the cosmopolitical project (Falk, 1995: 17).

In the sphere of political representation, some have proposed a world parliament on the model of the European parliament, and the Italian Peace Association has organized world assemblies, inviting representatives of peoples as opposed to states. As far as individual duties are concerned, the statute of the International Criminal Court has now been approved; if it is effectively instituted, it will at last allow us to judge the perpetrators of crimes against humanity. Progress is unbearably slow, but political institutions too must adjust to the boom of globalization. Why shouldn't the march of democracy – which has had to overcome a thousand obstacles to advance inside states – assert itself beyond frontiers, when every other aspect of human life today, from economy to culture, from sport to social life, has a global dimension?

Humanitarian interference

The model of cosmopolitical democracy summed up here has direct implications when it comes to policy-making. In which cases is the international community entitled to interfere in the domestic affairs of other communities? Abuse of the environment is not yet an international crime but if large populations, or humanity at large, are thereby threatened it may be seen as such in future. The discussion here though is focused on issues which have already triggered intervention. How do the cosmopolitical principles set out here relate to contingent problems such as ethnic cleansing, the repression of peoples and the daily violation of human rights?

From what I have said above it is clear that the cosmopolitical project is opposed to the stubborn defence of the traditional category of state sovereignty. Immanuel Kant noted that peoples had reached such a degree of association that 'a violation of rights in *one* part of the world is felt *everywhere*' (Kant, 1991: 107–8). You only have to open the morning papers to find detailed reports of infringements of accepted basic human rights somewhere in the world. International human rights protection devices can only respond as a tiny drop in the boundless ocean of abuses of power committed by, or with the consent of, states' governments. In such a situation, intervention is too precious a concept to be improvised or, worse still, used to disguise interests and delusions of grandeur.

During the NATO air raids on Serbia in 1999, Tony Blair, the most adamant supporter of the 'humanitarian' war, claimed that, 'it's right for the international community to use military force to prevent genocide and protect human rights, even if it entails a violation of national sovereignty'. Yet his argument – a veritable war plan for the post-cold war era – says nothing about which authority is legitimated to violate state sovereignty, who must suffer the use of military force or which human rights have to be protected. This is not an isolated coincidence: the more we read the statements of politicians and ideologists who periodically support war against states in which human rights are violated, the more we realize that no well-thought-out philosophy exists to guide the international community, invariably spearheaded on such occasions by liberal-democratic countries. On the one hand, war technologies have considerably increased their accuracy, with 'intelligent' missiles which now have a margin of error of just a few metres; on the other, there is a total short-sightedness about the political objectives to be achieved by war (see Kaldor, 1998). Ten years after the fall of the Berlin Wall, the 'baroque' category of sovereignty risks being replaced by something even more archaic: the law of the survival of the fittest.

The cosmopolitical perspective, instead, is informed by fundamental principles of tolerance, legitimacy and effectiveness. Tolerance serves to frame the violations of law correctly in political and anthropological perspective. The history of the human race is marked by amazement at the customs of other civilizations. Europeans have been at once the champions of anthropology, studying the habits of other populations, and ferocious oppressors of customs different from their own. During the Renaissance, seeing the pre-Columbian peoples' custom of making human sacrifices, the Spanish felt justified in committing genocide; and, in the very same years, the plazas of Spain were full of bonfires on which alleged witches were burnt. It is also fair to say, however, that in European society tolerance was born, and it is an antibody against the virus of genocide that Europeans can be proud of in so far as it developed much less fully and much later in other civilizations. Hence the ferocity of the Conquistadors was, albeit minimally, set off by the cries of outrage of observers such as Bartolomé de Las Casas and many others.

Far from demonizing 'otherness', the cosmopolitical perspective thus seeks to understand the reasons behind conflict and apply a positive adjustment to avoid something which could not be further from the principle of global responsibility: namely the policy of bias. In the second place, at moments in which the international community decides to interfere in situations which are under the jurisdiction of a given state, it is necessary to establish a precise hierarchy of instruments. It is one thing to have recourse to economic or cultural sanctions (as was the case against the system of apartheid in South Africa), another to resort to air raids. As things stand, the category of humanitarian interference includes under the same umbrella instruments that, from both the juridical and the political point of view, are totally different.

Military force must be used only as an extreme measure and only on the basis of a recognized international legality. There is no guarantee that an illegally promoted intervention will trigger anything good. By legality, we mean, first and foremost, the application of existing procedures, those envisaged, for example, by Chapter VII of the United Nations Charter. These procedures are by no means functional and may certainly be altered, but it would be unjustifiable to rewrite them unilaterally according to convenience.

Existing norms, however, reveal themselves to be totally incapable of guiding action in the eventuality of rights being violated inside a sovereign state. In this case – the most frequent in the last ten years – it is necessary to find meta-state institutions to legitimize the interference. The cosmopolitical model proposes that they must be founded on the world's citizens and their associations. Only in this way is it possible to prevent the slogan 'humanitarian interference' being used as a cover for what is in reality geopolitical interests.

The cosmopolitical proposal undoubtedly contains a contradiction. On the one hand, it delegates to structures devoid of coercive powers (such as international judicial bodies or the institutions of the world's citizens) the job of establishing when force must be used; on the other, it asks states – the sole depositories of coercive tools – to make their armed branch available to cosmopolitical institutions. But if the governments that defined themselves as 'enlightened' during the Gulf and Kosovo wars effectively intend to perform their democratic mandate, they should consult global civil society and international judicial authorities before they start flexing their muscles.

A humanitarian interference inspired by cosmopolitical principles should also rigorously separate the responsibilities of rulers from those of the ruled, especially when force is used. Once right/duty of a subject to intervene in another community has been accepted, it is still intolerable to believe that all the members of that community should be indiscriminately subjected to sanctions. If humanitarian interference is justified as an operation of 'international policing', then it is necessary to espouse fully the principle that distinguishes policing operations from military operations; the principle, that is, of protecting individuals and minimizing so-called 'collateral damage'. A democratic order is founded on the premise that sanctions should hit only those who have effectively violated the law.

Adam Smith expressed a concept that still permeates traditional foreign policies:

> If a government commits any offence against a neighbouring sovereign or subject, and its own people continue to support and protect it … they thereby become accessory and liable to punishment along with it … In a like manner a nation must either allow itself to be liable for the damages, or give up the government altogether.
>
> (Smith, 1978: 547)

On the basis of this principle, during the Gulf and Kosovo wars, the international community felt authorized to repress the Iraqi and Serbian people on account of actions directed by Saddam Hussein and Slobodan Milosevic.

In the cosmopolitan perspective, on the contrary, the citizens of an autocratic country whose government performs unlawful actions ought to be treated on a par with the hostages in a kidnapping. The use of force should guarantee the security of the citizens of the enemy country as well. What is striking about the interventions in Iraq in 1991 and in Serbia in 1999 is the total lack of relationship between the culprits of crimes and the individuals who suffered the sanctions. Saddam Hussein and Slobodan Milosevic are more firmly in power than ever, but new suffering has been inflicted on millions of individuals in Iraq and Serbia.

Much like a policing operation, 'humanitarian intervention' is judged by its effectiveness: that is, by its capacity to save victims and take the presumed criminals to justice. These are the criteria of effectiveness that ought to be borne in mind before and during the operations in question. The principles set out here are clearly different from the ones which inspired the Gulf war in 1991 and the 'humanitarian' intervention in Kosovo in 1999. In both cases, the international alliance guided by democratic states resorted to the use of military force long before other instruments, such as diplomacy and sanctions, had completed their course. The cosmopolitan deontology that I propose here would have envisaged a very different *modus operandi*, placing the onus on the civilian populations who are the first victims of war. It would have offered a prospect of economic and social development founded on the integration among societies, thus depriving the warmongers of mercenary arms and social consent. It would have asked the peoples in question to turn against dictators who spoke of ethnic cleansing or the annexing of other states. It would have risked placing huge numbers of blue helmets on the ground, accompanied by representatives of civil society and peace workers (the so-called 'white elements').

It is difficult to say whether the means proposed by the cosmopolitical deontology would have proved effective in achieving specific aims, such as the restoration of sovereignty to Kuwait and a timely end to the genocide in Kosovo. But one only has to see the results of interventionism based solely on bombing to realize that the international community's cure was much worse than the sickness. Almost a decade after the Gulf war, Saddam Hussein is still in power in a country on its knees on account of his dictatorship and the economic embargo. In Serbia, Milosevic continues in power with some local support. In Kosovo, ethnic cleansing continues. The only features that have changed are the people on the receiving end and the directions in which the hordes of refugees are walking. Several peoples have been returned to a pre-industrial state. This is not the cosmopolitical responsibility we are fighting for.

Note

1. Ambitious proposals to reform the world order have been formulated by the Commission on Global Governance (1995). On the issue of democratization, the former Secretary-General of the UN, Boutrous-Ghali (1996), has released a specific *Agenda*, which, unfortunately, received much less attention than his previous *Agenda for Peace* (1992). The relationship between the UN and democracy is discussed in Archibugi and Held (1995) and Archibugi, Balduini and Donati (2000).

14
An International Court of the Environment

Amedeo Postiglione

Introduction

It is well understood today that the 'environment' is an issue of global scope. Even apparently local disputes between or within nations have global consequences. Ultimately local exploitation, under the existing economic regime, may result in the exhaustion of the natural foundation for the future economy (see Daly, 1996). Accordingly new institutional arrangements at global level are needed to resolve these disputes authoritatively in a way which takes account of the need for justice to future generations as well as justice to a wide range of parties affected in the present (see for example Singh, 1985; Sand, 1991; Palmer, 1992; Stone, 1993).

One important possibility explored in this chapter is an International Court of the Environment. The nature of international environmental disputes is first explored. Then the function of an international jurisdiction for the environment is considered and the conditions necessary to support such an institution are set out and compared with those supporting the International Court of Justice at The Hague. The functions of negotiated settlement and arbitration are compared with that of adjudication which, it is argued, is indispensable in the current circumstances. In conclusion, the International Court of the Environment Foundation supports gradual steps to reform UN institutions to establish the focus for a specific sphere of environmental jurisprudence complementary to but necessary for more effective negotiation and arbitration.

The nature of international environmental disputes

Many current environmental disputes are global in scope, interdependent, indivisible, and future-oriented (see French, 1992; Timoshenko,

1992). They raise issues of equity and transparency. They produce wide-ranging and long-term effects on the overall ecological system (the global environment) and, therefore, on the ability of life to sustain itself on earth. Such phenomena are the reduction of the ozone layer (ozone depletion), climate change, the loss of biodiversity, desertification, deforestation, soil erosion, salinization, acid rain, radioactivity in the air, soil, and in rivers, lakes and seas, pollution of rivers, seas and the atmosphere, genetic manipulation, and individual ecological disasters such as Chernobyl, *Exxon Valdez*, *Amoco Cadiz*, Bophal, and Seveso (see Low and Gleeson, 1998; Low, 1999).

Damage done to a natural resource sooner or later becomes damage done to the environment ('nature') in general. Damage done to the environment is damage done to humanity and to society. In every case of environmental damage there is always a social dimension. As of today, future generations must be assured of the right to life (Abravanel, 1995). Indeed this is the first time in human history that the problem of the adaptation of legal and institutional systems to humanity as a whole and to its future has been posed. It is just the most general and universal part of environmental damage that remains as yet hidden, owing to the working behind the scenes of powerful political and economic interests that have no interest in transparency. The weight of new phenomena must also be assessed for their importance, for example demographic conflicts, conflicts arising from overpopulation, ethnic conflicts sharpened by environmental degradation, conflicts due to deprivation (e.g. of agricultural land), migration conflicts, imperialistic ecological exploitation, water deviation projects (e.g. the Rhine, Euphrates, Tigris, Nile), socio-ecological conflicts (viz., oil exploration, nuclear testing, transport of hazardous waste, industrial dumping, overfishing), acts that the world even now judges to be environmental crimes (the setting on fire of the oil wells of Kuwait during the Gulf War), and armed international conflicts (see Homer-Dixon, Boutwell and Rathjens, 1993).

Synergetic and cumulative phenomena – if the pollution-oriented economic model of production and mass consumption by the more developed countries remains substantially unchanged – will surely mean the worsening of the global ecological crisis. Clearly, then, realistic proposals for the reform of the current models governing the resolution of environmental disputes internationally must take account of the above characteristics of these disputes (International Court of the Environment Foundation, 1992; Kimball, 1992).

The function of an international jurisdiction for the environment

In principle, the creation at the global level of a Court for the Environment is justified just because of the objective characteristics of the serious environmental disputes that have already taken place, and of the many more numerous potential ones (Bourgeois, 1992).

Jurisdiction does not mean only 'the binding ascertainment of law' (a concept which is also present in voluntary arbitration) but also the power to apply international law on the environment, with authority and effectiveness, by independent judges who are the expression of a supranational authority. The decision of the court must also hold in the event of disagreement on the part of a nation-state, or states, which are parties to the dispute. Jurisdiction (from the Latin *jus dicere*) does not imply only a declarative power, but also the exercise of an *imperium*, which is to say, of a *binding power that enforces the law*. This is a power that does not arise out of the will of the parties but from a legal authority belonging to a superior order that is able to impose its will even against the will of the parties themselves, or against their simple inertia.

This formulation, typical of a globalist conception, implies the surrender by states, within certain limits, of some of their sovereign prerogatives (see Marchand, 1992). Even though it is foreign to the current phase of development of international law, it is also the most practical and functional means for the resolution of the global environmental problems described above.

In order for the project for an effective international environmental jurisdiction to take concrete form, a number of premises must exist:

- a new legal base must be created, that is, a framework convention of the states (or at least of the majority of them) must be drafted;
- a court must be created as a supranational authority, with decision-making powers that are effective *erga omnes*;
- the Court must be accessible to individuals and non-government organizations, and not just to states (since the Court would be the expression of the international community, and since the necessity for the enforcement of the human right to a healthy environment would be involved);
- a number of other elements too would have to be present, namely, a body of judges that is not only independent but has its own status of non-removability. These judges would have to be specialized and required to comply with the principles of international law, but within the application of environmental law. Hearings would have

to be normally conducted in public; decisions of the Court would be founded on specified grounds. The Court would have the power to carry out controls related to prevention on an *ex officio* basis, the power to adopt emergency measures, including temporary or restrictive injunctions, the power to order economic sanctions such as an injunction to restore the environment to its original state, and, subordinately, to compensate for damage.

The desired international environmental jurisdiction is too far from the existing model of the International Court of Justice at The Hague, even after the creation of a specialized division for the environment (which has already taken place). This is because the fundamental legal and institutional premises conditioning the practical functioning of such a jurisdiction are missing. Moreover the legal basis of the current International Court is wholly inadequate, since it is still constituted by a Statute annexed to the United Nations Charter, itself the expression of an age that is by now past and gone (Bilderbeek, 1992). The International Court, looking beyond the words 'Court of Justice' is substantially a court of arbitration, since it judges on the basis only of the agreement of the parties to the dispute. Access to its justice is reserved only to states, with the consequence that its role of administrator of *ecological* justice in the world is practically ineffective.

It should also be remembered that the Court of Justice at The Hague is recognized by fewer than one-third of the states in the world (49 out of 180), that some countries, such as France (cited before it in 1973 owing to its nuclear tests in the Pacific) and the United States (after its intervention in Nicaragua in 1984) have actually *revoked* their acceptance of it. It should also be recalled that no suit regarding the environment has ever been presented before it since the Second World War, so that no jurisprudence exists in the environment sector.

With all due respect for this institution, it would appear unrealistic to expect the International Court of Justice at The Hague to provide an international jurisdiction for the environment without its radical reform by means of a new Convention. Indeed it would be more correct to speak of its *refounding*, since in fact it has never existed as an environmental jurisdiction.

A variety of contemporary and future shifts will instigate a very desirable reform of the current United Nations juridical model. These changes include the speeding up of the tempo and the deepening of the ecological crisis, the multiplying of environmental disputes over damage done to the environment and over the inequitable distribution of common resources, and social pressure for real access to ecological justice. Other forces for change are evident on a global scale, including the evolution of international law on the environment in terms of greater completeness, harmonization between sectors, definition of the obligations of the states, and creation of more suitable instruments for its application.

Not to be excluded, of course, is the possibility of a new role for the United Nations in the resolution of transnational environmental disputes, both administratively (prevention and management) and jurisdictionally (ascertainment of legal responsibility and application of sanctions where called for). Included in such a possibility is reform of the International Court of Justice at The Hague, as an organ of the United Nations. However we believe another opinion to be legitimate.

It would appear that the environment, owing to its special characteristics, would be better served by setting up an autonomous International Court for the Environment, as a specialized body composed of a few independent judges who are competent to deal with this complex subject, accessible not just to states. This new institution should come into being on the initiative of one of the states, on the basis of a Framework Convention made up by the states who perceive the necessity for and the urgency of its creation. It could exercise gradual powers, working up a unitary jurisprudence on the basis of principles common to the various sectors (climate, oceans, forests, etc.) thus avoiding the jurisdiction being broken down into various separate authorities. This is the proposal that the International Court of the Environment Foundation (ICEF) has been putting forward for some years now (see Exell Pirro, 1992; Postiglione, 1993, 1994, 1995).

During the UN Conference on Environment and Development (UNCED) held in Rio de Janeiro in 1992 nothing new was decided as to a credible model to govern the protection of the environment internationally (except for the Commission for Sustainable Development). It must also be acknowledged that the very existence of the International Court of Justice at The Hague is criticized by some as a mere pretext for leaving things as they are, despite the worsening of the global ecological crisis.

Since the economic and political difficulties are real, courage must be found to discuss them objectively and realistically. Therefore, ICEF appeals directly to governments regarding this new initiative, which will open up the way for positive discussion and lead to some important results early in the twenty-first century. One possibility is to form at once an International Court of the Environment on a voluntary basis by some governments, to start off a gradual process of experiment.

The concept involved is of a forum that declares what existing international law on the environment is, as the first step towards the final objective of a new Court.

The function of international arbitration in environmental disputes

Substantial room has already been made for arbitration in the resolution of (non-environmental) disputes in the various legal orders internal to the states and in the international legal order itself (the Permanent Court of Arbitration; the International Court of Justice at The Hague). Clearly, environmental matters too can make use of alternative models for the resolution of disputes that bypass the proceedings of ordinary courts, and therefore the ordinary operation of adjudication (Jonkman, 1994).

The simplest form is *negotiated settlement* effected by the parties themselves, an activity that may also have a function preliminary to or having effect at the same time as *arbitration* (characterized by voluntary submission to the decisions of a disinterested outside party) or as the proceedings of ordinary courts of justice. Resort to alternative models for the resolution of disputes (negotiation and arbitration) in the various domestic legal orders is made possible by private and public initiatives, since it is faster, less cumbersome and more practical.

Having acknowledged the positive, and desirable, role that arbitration can play in resolving environmental disputes both nationally and internationally, the problem then arises as to its relationship with *adjudication*. It would appear best to underscore a few aspects, in order to be able to discuss the question with greater clarity:

- In national legal orders arbitration has a complementary function, since resort to the courts exists as a certain and authoritative possibility within the system. But this is not the case at the international level. In fact, in order to give international arbitration a positive role in relation to environmental matters it is necessary to give the global community a true adjudicating body, that is, to give it an International Court of the Environment (as a new specialized institution or as the utilization of the International Court of Justice at The Hague, but radically reformed).
- An International Court of the Environment will be of its nature a permanent public institution, while arbitration maintains its private and voluntary nature. It will be founded on the willingness of the parties to the individual dispute to submit to it (also assuming its

institutionalization, as is the case of the Permanent Court of Arbitration at The Hague, which is only a periodically updated list of judges from which the states may choose, to settle their disputes, while the procedural rules are kept to a minimum and may even be wholly set aside by the parties).

- The efficacy of the arbitrator's decision is limited to the parties (questions of mutual or bilateral interest), while the decisions of an International Court of the Environment may have efficacy *erga omnes* where it has to deal with world-spanning problems involving many parties, such as those who cause damage or are the victims of the damage. It is perfectly clear that since many environmental disputes are global, indivisible, interdependent, and not able to be separated into parts because of their objective supranational nature, there is no room for arbitration. In fact, any arbitration decision could solve only a portion of the dispute, but not the dispute as a whole.

- The problem of access to justice in environmental matters is typical of a court, and here there are as yet obstacles at the community, national and international level. In arbitration, access has to do only with the voluntary choice of convenience made by the parties to the dispute. Therefore it is necessary to promote the Court for the Environment by connecting it with the social and democratic exercise of every individual's human right to the environment in the spatial sense. It does not make sense to grant access to justice within a state on the part of a person or non-governmental organization but deny it conceptually and legally in the international sphere in cases where the dispute goes beyond the bounds of a state's jurisdiction.

- The court provides a certain guarantee of public notice and of transparency, which is important for environmental questions, notwithstanding the desire of the parties for a climate of confidentiality and even secrecy. This aspect cannot be ignored with reference to the right to environmental information and to the right to participate.

- The environmental court will create jurisprudence that is often innovative and evolutionary, and will enable progress to be made in environmental law. This phenomenon has been noted favourably in every state, and as regards the work of higher courts, which often play a propulsive role, such as Italy's Constitutional Court and Court of Cassation. It is a positive fact recorded both in common law systems and in other legal orders, where – quite beyond appearances – the court does not limit itself to literal interpretation of the existing body of law in sectoral fashion, but has successfully attempted the systematic elaboration of common principles.

The possibility of international arbitration on environmental issues evolving procedures and principles of law is not to be excluded, but appears more limited. It is true, however, that repeated and authoritative opinions can also create an *opinio juris* that is legally substantive.

Considering that the European Commission Court of Justice, with its headquarters in Luxembourg, has created the principle of the supremacy of Community law (law of the European Union) over the domestic law of the member states, we can understand the importance the proposal for an International Court for the Environment may have for the effective evolution of international law and its effective application. It would provide for supremacy over the legal systems of the individual states in the global community. Not only could treaty provisions receive better enforcement, but so would principles of customary law, among which are the *pacta sunt servanda* and liability for massive damage arising from transfrontier pollution falling in other states or in areas outside of any court jurisdiction.

An International Court of the Environment would favour the current trends toward the integration and harmonization of the various legal orders. It would do this by ensuring not just the regulation, but also the agencies necessary to their enforcement.

The International Court of the Environment would have a preventive role (emergency measures) that seems less consonant with the role of arbitration, since arbitration presupposes a firm dispute, one defined in its own terms, not an open and dynamic conflict whose potential effects could lead to further damage. The court would provide guarantees of the judges' independence and impartiality, their being called on to be disinterested in the outcome of the dispute, not just with reference to two or more parties, but with respect to all the interested parties: the victims and those who have suffered monetary damage.

As a consequence, the International Court of the Environment appears more suited to dealing with that part of environmental damage that remains hidden, however real and serious it may be, *behind* the subject of the quarrel submitted to arbitrators. Such a quarrel may be defined by the interests only of the parties to the dispute.

Conclusion

The ICEF believes it important to go ahead with action promoting a true court of the environment at the global level. ICEF does not take its inspiration from an institutional idea of environmental protection that is purely horizontal, acting at the level of governments and entirely within the current structure of the United Nations. It deems such structure unsuited to the aim, but rather takes inspiration from a globalist idea. Such an idea postulates the need for a higher level of cooperation among states. This new cooperation would imply the surrender of certain prerogatives of national sovereignty, through the instrument of a new Convention that will create a global authority for the environment, broken down into an administrative authority and a juridical authority.

The extraordinary seriousness, complexity and globality of the phenomena involved in the deterioration of environmental systems and the acceleration of the ecological crisis demand clear and responsible choices, since what is at stake is the capacity of life to sustain itself on earth. The retention of this capacity is a supreme value that is not subject to the disposal of any government. Therefore, the right and duty of every person or non-government organization to demand democratically and forcefully a different model of environmental protection at the global level must be acknowledged and respected. In the meanwhile, every gradual step towards this objective is to be greeted with favour, provided that it is not exploited to combat the choices necessary for the good of future generations.

It is not to be wondered at then that ICEF supports, faithfully and clearly, present efforts to update and upgrade UN institutions that are to be found within the UN and the bodies connected to it. Examples of these efforts include the creation of the specialized environmental section of the International Court of Justice at The Hague, the attempt to find an environmental role for the permanent Court of Arbitration, and the initiatives aimed at creating a new International Court of Conciliation and Arbitration for the environment. Every important and businesslike initiative aiming at the good of the environment at the global level is to be encouraged in a spirit of mutual respect and cooperation, since there are many ways which must lead to overcoming the powerful wall of economic and political interest opposing the necessary reforms.

Peace, development and environment are interdependent and indivisible in a world worthy of humanity, since they form the fundamental principles for ensuring the capacity of human life to sustain itself on earth. Today, it is no longer enough to affirm these principles, but efforts must be made to find the tools for their urgent and concrete application.

An international court and an international board of arbitration for the environment, properly understood, are complementary and can play an important role, if they have as their frame of reference a common strategy. Without a true court at the global level, arbitration itself will have no institutional base of reference, nor any true possibility of developing.

The call for an International Court of the Environment is justified not only by its human rights aspect but also by the strongly felt social and ethical need for environmental justice. There is a close link between environmental and social problems, and international environmental law cannot fail to be its interpreter.

It should not be forgotten that the main source of injustice is also the monopolist and plunderer of the South's natural resources. A proper interpretation of human rights focuses firstly on the recognition of civil and political rights and then on economic, social and cultural rights (equal pay, health, education), and finally on upholding the rights to solidarity (the right to peace, the right to self-determination, and the right to the environment).

This evolution of human rights, which is equivalent to an enrichment of the value of human dignity on a universal basis with all its potentiality, is also a painful struggle for freedom, a rejection of every kind of violence, a need for justice, peace and solidarity. Fighting for human rights has become synonymous with fighting for a new society. Therefore, the serious and complex ecological issue that is characterized today by global pollution and an inequitable model of production and consumption is linked with social issues throughout the world. Conciliation and arbitration, although very useful instruments, may not be sufficient for justice as they make the solution of disputes dependent on the consent of the states. However, it is commonly known that the states themselves may commit environmental crimes or tolerate the presence of perpetrators of such crimes within their territory. It therefore seems advisable and realistic to work towards strengthening international judicial guarantees for effectively protecting the human right to the environment. Above and beyond its legal aspects, an International Court of the Environment is therefore claimed to be a future force acting as guarantor of the integrity of life in the future.

15
Humane Governance and the Environment: Overcoming Neo-Liberalism

Richard Falk

> To blow this great blue, white, green planet, or to be blown
> from it.
>
> Saul Bellow (1970: 51)

Introduction

The idea of humane global governance is based on both functional and
normative considerations (Commission on Global Governance, 1995;
Falk, 1995). The functional dimension of global governance responds
to the complexity and interrelatedness of many dimensions of social,
economic, cultural and political life, and the practical importance of
securing reliable and beneficial transnational arrangements.

The normative dimension of global governance is preoccupied with
the fairness, sustainability and democratic quality of these arrangements, and accounts for the adjective 'humane'. At issue, in part, is
whether humane governance is achievable within the prevailing
framework of ideas operative in the world today, especially ideas associated with property rights, market relations, nationalism and sovereignty. The argument of this chapter is that such ideas must be
modified, both in their interpretation and application, if the project to
establish a world order based on humane governance is to become a
politically relevant undertaking, and not just an empty exercise in
wishful thinking. Global governance does not imply any particular
degree of centralization, and 'governance' must be sharply distinguished from 'government', which does imply a centralized institutional arrangement as the basis of authority and order (Rosenau and
Czempiel, 1992).

Governance seeks effectiveness and legitimacy of political arrangement in a flexible manner that encompasses networks, informal regimes, and customary linkages, with a minimum degree of bureaucratic centralism and a maximum amount of political space for exploration and diversity. As well expressed by Marie-Claude Smouts,

> the concept of governance presents numerous advantages: it is flexible, adaptable, it takes nothing for granted; it encompasses a great diversity of actors and describes an ongoing process of interaction that is constantly changing in response to changing circumstances; it denotes a form of social coordination which can take into consideration various public and private interests in the management of matters of common concern and which takes responsibility for these matters collectively.
>
> (1998: 295)

As has been well observed by Michael Anderson, '[L]ike human rights, environmental law houses a hidden imperial ambition; both potentially touch on all spheres of human activity, and claim to override or trump other considerations' (Anderson, 1996: 1). Safeguarding the environment is, of course, an essential part of the project to establish humane governance, but it is only one aspect of a far wider undertaking that involves social justice, democratization and demilitarization. And even within the sphere of environmental protection one can readily imagine, and even anticipate, an effective approach that is unacceptably 'inhumane'. For instance, one could easily believe in prospects for the emergence of an authoritarian control system imposed on the world so as to preserve unequal access to resources in the face of intensifying scarcities and environmental decay. The essence of a humane approach is the assurance that all peoples have their individual and collective rights realized, including economic and social rights that ensure the right to life, to subsistence, and to an international order capable of ensuring other rights. If not yet attained, then at the very least, humane governance insists that such rights are being actively affirmed as policy goals to be seriously pursued.

The difficulty of achieving humane governance for the environment is complex and controversial. The deepest root of the difficulty arises from the interplay between the political fragmentation of sovereign territorial states governed by sole reference to 'national interests' and the variously bounded ecosystemic realities that make up our biophysical world. But currently the most environmentally dangerous circumstance arises from a drive to unify the world economy on the basis of profit-maximizing criteria that continuously stimulate consumerist demand of a rising world population for an environmentally destructive life-style. It is impossible to envisage much movement towards humane governance without some prior transformational process reducing the impact of these economistic forces.

The reduction of economism is now such a high priority that it prompts a revisionist view of the role of the state in improving the quality of global environmental governance. Only the state, among existing political actors, has the potential capabilities to implement a degree of environmental regulation that will be needed to safeguard the health and wellbeing of peoples now alive on the planet and dis-charge responsibilities to future generations. But the advocacy of such a re-empowerment of the state for the sake of humane global gover-nance is not meant to restore a world order in which the state becomes once again autonomous, operating on the global stage as the only legitimate political actor, the so-called 'Westphalia system'. Those times, happily, are past (Falk, 1998).

A re-empowered state would act alongside other political actors, including those representing civil society, and have as its most urgent mission the negotiation of a global social contract with market forces that would include environmental protection as a vital element. The quality of this environmental protection would combine notions of sustainability with ideas of equity so as to offset the implications of poverty and resource disadvantages to the extent possible. It is relevant to take note of the recent massive extension of coastal state authority in relation to the ocean, through the establishment of a 200-mile Exclusive Economic Zone that confers rights over resources upon states, but also imposes responsibilities for environmental protection. Such a statist approach to environmental protection is a contribution to global environmental governance that goes against the wider per-ception of the decline in the role of state.

Points of departure

This chapter explores the ethical concerns raised by the impact of glob-alization upon the overall effort to respond effectively to the various dimensions of the environmental challenge. It proceeds from the gen-eral proposition that globalization is dominated by a coherent set of ideas, often labelled as 'neo-liberalism' or 'the Washington consensus', that are antagonistic to, and tend to override, environmental concerns.

The dynamics of global economic governance push further in this direction, involving a dedicated effort by business and financial interests to make the world economy work as efficiently as possible as assessed by aggregate figures reporting economic growth and profits. Given such priorities, the intrusion of environmental protection is perceived as both a burden on capital formation and an unwanted interference in the market.

These ideas are strong, and remain dominant, in both the discourse about globalization and the shaping of policy, putting environmental concerns on the defensive. Part of the explanation arises from the extent to which governments representing states have accepted as their own outlook the framework of ideas embedded in neo-liberalism and have acted accordingly, jeopardizing territorial interests in the process, including the interest in a clean and healthy environment. In this regard, it is important to realize the extent to which states have become *instruments* of the private sector, including its transnational outlook, with a loss of capacity and will to promote the *public good* in general, and its environmental aspects in particular.

This overall set of circumstances provides a further occasion for worry due to the low level of institutionalization of environmental protective activity on regional and global levels. It certainly makes as much, if not more, sense to create a capacity for global governance in the environmental sector as it does with respect to trade and development, but the ideological climate is not so inclined. The World Trade Organization and the United Nations Development Program (UNDP) reflect the effects of functional and normative pressures with respect to trade and development concerns about governance. The UNDP has itself been severely weakened in recent years to the extent that its undertakings are perceived as being guided more by concerns of developing countries, that is, with people and poverty, than with growth and profits. The United Nations Environmental Program (UNEP), operating on a tiny budget, located in Nairobi outside the main policy-making centres for global issues of Geneva and New York, should not even be taken seriously as a participant in environmental governance. The main useful role of UNEP is informational, and at most advisory, but the scale of its activity is so small relative to the scope of the environmental challenge, that it plays almost no role in serious global efforts relating to the environment.

What have existed, and could conceivably have served as a prelude to the establishment of environmental governance under UN authority, have been well-planned conferences under United Nations auspices, initially at Stockholm in 1972, and then again at Rio de Janeiro in 1992. These events have provided arenas for transnational social forces deeply distressed about environmental deterioration and the absence of environmental governance, to interact with governments, and to put forward their case for a less market-oriented approach to global policy.

Whether these conferences have been a success is subject to debate, but they have at least raised environmental consciousness at the level of the state to much higher levels and have engaged popular consciousness and the media, at least for short intervals. But their contributions should not be exaggerated. Their influence tends to diminish over time, and despite modest efforts to avoid this outcome, such as the establishment at Rio of a new body, which meets periodically and is called the Commission on Sustainable Development, the evaporation of media concern and the dilution of a policy focus on global environmental policy is evident. Additionally, large governments and corporations felt threatened by the buildup of pressures of this sort, and seem intent on preventing UN sponsorship of large consciousness-raising conferences on contested global issues.

Perhaps of greatest practical relevance have been regional and global law-making attempts to bring into being a treaty regime capable of responding to those aspects of the environmental challenge that possess a global scope. The Ozone Depletion Treaty and Protocols have been taken by optimistic commentators as expressive of confidence that when global environmental problems become really serious, appropriate, constructive action will ensue.

In effect, the feedback mechanisms are adequate to the challenge, given the rationality of public opinion and leaders, and nothing more by way of governance is required. The problems associated with ozone depletion have certain helpful features: the scientific assessment of cause and cure was widely accepted, commercial substitutes for ozone-depleting CFCs existed and had been developed by the main producers of CFCs, the adjustment costs arising from a shift to more benign chemicals for refrigeration and aerosols were not too large, and the richer countries were sufficiently concerned about the effects of further depletion that they were prepared to bear the financial cost of adjustment for developing countries.

But global warming is quite a different story. The scientific community is somewhat divided on diagnosis and cure, the adjustment costs are huge, the richer countries are reluctant to pay these costs for their own adjustment and resistant to bearing the entire burden for poorer countries. The effort in 1997 to produce a law-making treaty at Kyoto did generate a Kyoto Protocol that placed a schedule of graduated limits on the emission of greenhouse gases. This was widely endorsed at the time by most governments, but it has not been followed up either by shifts in behaviour and regulation by states of their own societies or through ratification of the global framework of constraints embodied in the Protocol. As in so many areas of global policy significance, it has been the United States that has been the most obstructive political actor in relation to fashioning a responsible approach to global warming and to the environmental challenge generally.

The governance potential of the lawmaking approach can best be appreciated in relation to the great achievement of international bargaining that produced the UN Convention on the Law of the Seas in 1982. Here a comprehensive framework of legal norms, including innovative dispute settlement machinery, was agreed upon after a decade of effort in a comprehensive text that can reasonably be viewed as establishing a public order of the oceans in many respects responsive to the challenges of the times. There are major shortcomings, including an excessive reliance on the capacities and rationality of coastal states and an unfortunate delay in US adherence, but despite these setbacks, the path to humane governance without awaiting the establishment of governmental structures is impressively realized in relation to the oceans.

A more modest success of the same sort was achieved in relation to the governance of Antarctica. If a comparable comprehensive treaty regime for the global environmental challenge existed, it would dispel much of the pessimism now present with respect to dealing with the global environmental agenda in a manner compatible with the ethical precepts of human-centred development. There are circumstances under which states can use law to structure cooperative arrangements that work for both mutual benefit and the public good, and do so in a manner that does not directly affect sovereignty.

But global environmental policy, with some regional exceptions, has not been one such area. The protection of the environment is made difficult because it interferes in the workings of the market. It is also often complicated by its non-territorial locus that makes it almost impossible to assess blame in a manner that commands respect. Environmental challenges of global scope also generally emerge by reference to a long and elusive time dimension, making them hard to analyse until the threshold of irreversibility has been approached, if not crossed. A further complicating factor is demographic pressure on the demands for renewable resources. Poor countries are made especially vulnerable to environmental devastation due to such unsound practices as deforestation and insufficient freshwater resources.

The catastrophic impact in 1998 of Hurricane Mitch on Central America was increased by floods and mudslides, which would not have occurred to nearly the same degree if the forests had not been depleted by poor people needing firewood. Iraq's great civilian losses in the aftermath of the Gulf War were brought about in part by the unavailability of safe drinking water in relation to the demand. Thinking ahead, these pressures on renewable resources will certainly grow in the decades ahead. The present expectation is that by 2025, at the latest, the world population will grow by two billion, increasing to almost eight billion, with more than 90 per cent of this increase being concentrated in developing countries that are already confronting water/forest shortages. Such problems of meeting urgent needs in conditions of acute scarcity are almost certain to divert attention from seemingly more remote challenges arising from the deterioration of more 'distant' regional and global environmental conditions.

There is a final dimension of difficulty, which arises from the altered orientation and capacity of states, which remain the dominant political actors with respect to the formation and implementation of global environmental policy. The leaders and policy-makers, aside from their collaboration and frequent identification with global market forces out of varying mixtures of conviction and opportunism, are generally guided by short-term time horizons of accountability, whereas environmental threats associated with global conditions are of a far longer duration. The temptation is almost irresistible to transfer the adjustment costs to future leaders rather than summon the effort to respond adequately in the present setting. The 'realism' or Machiavellian world picture that tends to guide most political elites reinforces this orientation. This outlook regards ethical factors associated with wellbeing and improvement of the human condition as irrelevant, even inappropriate, considerations when it comes to shaping global policy. The role of ethical factors is supposed to be operative to some extent in state/society relations *within* the state, although even here the neo-liberal mentality works against the idea of the moral or compassionate state, preferring instead the efficient and cruel state. For global action, that is, undertaken *outside* the state, there is little acceptance of the existence of the global public good, which it is the duty of governments to sustain.

This overall situation is accentuated in the present period due to the geopolitical style of leadership provided to the world by the United

States Government. Washington has championed a realist/neo-liberal outlook that falls short of what even many other governments in advanced industrial and post-industrial countries are willing to support in relation to global environmental responses. As such, it opposes both ethical and environmental claims if their effect is to interfere with market forces or with American foreign economic interests in investment or growth.

At the very historical moment, following closely on the end of the cold war, that many people were hopeful that conditions finally existed to create an effective and creative United Nations, the Organization is being bypassed in relation to the most important global policy issues ever relating to peace and security. The reliance on NATO to carry out Western policy in relation to Bosnia and Kosovo is a manifestation of this downgrading of the UN role, but there are many others. The reluctance of major states to allow serious discussion of disarmament or environmental governance within the confines of the United Nations is indicative of this mood. This mood opposes, on principle, initiatives that rely on institutional regulation or that threaten to hamper the operation of so-called self-organizing systems, of which the market and the Internet are the prime examples.

First steps towards humane governance of the environment: the emergence of normative ideas

Normative ideas about the environment combine notions of ethics with those of law. These ideas reflect the realities of a decentralized world order system with power and authority widely dispersed, and as a result are enunciated as essentially voluntary guidelines unless incorporated in a more obligatory form in a specific treaty instrument. But as guidelines undergirded to some extent by the authority of international law, such norms are available for degrees of implementation through initiatives pursued by civil society associations. This process has been surprisingly effective in the sphere of human rights, and sporadically effective in relation to opposition on environmental grounds to specific practices: whaling, proposed mining in Antarctica, atmospheric and undersea testing of nuclear weaponry, ozone-depleting chemicals.

The environmental movement was born in the 1960s when concerns about the sustainability of advanced industrial civilization assumed popular prominence. The eloquent voice of Rachel Carson warning that the widespread use of DDT was threatening natural habitat to such an extent that birds would soon disappear from the planet exerted an immense influence on public awareness. Also, the sombre studies of the Club of Rome to the effect that population growth, pollution and resource depletion would soon induce civilizational collapse served as a wake-up call with respect to environmental protection (Carson, 1962; Meadows and others, 1972). A culminating occurrence in this period was the convening in Stockholm of the 1972 UN Conference on the Human Environment, with its important attempt to set forth a normative framework in the Stockholm Declaration (Weston and others, 1997: 866–70).

While a great contribution with respect to consciousness-raising and agenda-setting, Stockholm was also a scene of disappointment owing to the reluctance of the richer countries to push for a serious environmental agency to be established within the UN system, a divisive split between North and South on the relations between environmental protection and economic development, and an unwillingness by the United States in particular to grant the relevance of wartime operations to environmental protection. At the same time there were important contributions made. Many governments came to realize their own complacency with respect to environmental policy, and went about establishing environmental ministries and programmes. Civil society took good advantage of the Stockholm arena to push the intergovernmental process further and to gain access for their views in the global media whose representatives were gathered for the official conference.

To a significant degree both the problems and the contributions persist. The challenge of environmental protection, as earlier suggested, has been seriously aggravated by the hostility of the neo-liberal consensus to all forms of global regulation. But Stockholm also encouraged a learning process that was most effectively evident in the workings of the Brundtland Commission, and its final report, *Our Common Future* (Commission on Environment and Development, 1987). Brundtland helped greatly to build a consensus in support of reconciling North/South differences by considering environmental protection under the rubric of 'sustainable development' and by affirming the degree to which 'poverty' was a form of 'pollution'. As a matter of discourse, then, the preoccupation with environmental decay as such was linked, and verbally subordinated to, the imperative of economic development, and the plight of poor countries and of poor people. The Brundtland Commission report was efficiently distributed within the UN system, and provided a common background for participants in the Earth Summit held in 1992.

Aside from Agenda 21, which detailed the implementing action to be taken with accompanying budgetary estimates, the main normative outcome was the Rio Declaration on Environment and Development, as formally adopted at the end of the UN Conference on Environment and Development (Weston and others, 1997: 1112–15). It is a useful summary of normative thinking, but its mandates are far too general to be very helpful in specific settings. It was also evident at Rio that North/South tensions, although papered over, persisted, and that new North/North tensions were evident with respect to sharing the burdens of adjustment costs, and even in relation to whether global collective action was necessary at the present time. Rio was also notable for the efforts of its organizers to give transnational civic groups a role in the proceedings, and to give an even bigger place to business groups representing the transnational private sector.

The dominant normative ideas contained in the Rio Declaration are worthy of some specification as reflective of what states have agreed upon *rhetorically*, but by no stretch of the imagination has this agreement been acted upon *behaviourally*. These ideas can be listed as follows:

- sustainable development as the framework principle;
- permanent sovereignty over natural resources vested in states;
- state responsibility for environmental harm to others situated beyond territorial limits;
- right to development;
- equitable concern for the needs of present generations and fairness to future generations;
- cooperation among states to eliminate poverty, reduce disparities in living standards, and meet the needs of the peoples of the world;
- recognition of special needs of developing countries, especially those that are least developed and most environmentally vulnerable;
- acknowledgement by developed countries of their greater responsibility to bear adjustment costs associated with protecting and restoring 'the health and integrity of the Earth's ecosystem';
- duty of states to reduce and eliminate 'unsustainable patterns of production and consumption';
- duty of states to 'promote appropriate demographic policies';
- an affirmation of a democratic approach to environmental policy based on participation, access to information, education, and transparency;

- a contextual approach to environmental management by states, reflecting the extent of the environmental challenge, but also appreciative of developmental pressures on poor countries;
- establishment of procedures at national and international levels to fix liability and compensation for environmental damage;
- the duty to prevent export of environmentally harmful activities and substances;
- the acceptance of the precautionary principle in circumstances where scientific certainty is absent, especially if serious irreversible environmental damage is threatened;
- the encouragement of the polluter pays principle;
- the duty to notify affected states of potential adverse environmental effects arising from natural disasters and activities;
- special recognition of the special role and needs of women, youth, indigenous and local communities, peoples living in oppressive circumstances, in relation to sustainable development;
- an affirmation that war 'is inherently destructive of sustainable development' and that states engaged in war should show respect for international law in relation to the environment;
- the obligation to resolve international environmental disputes by peaceful means;
- the duty of states to cooperate 'in a spirit of partnership' to realize these normative principles and to develop international law suitable for sustainable development.

Beyond the normative framework of the Rio Declaration there are several other normative ideas that have a potential bearing on environmental protection that deserve consideration, and enjoy a provisional status as obligatory features of international law:

- the general community duty to respect the integrity of the global commons, and to act responsibly in relation to oceans, space, polar regions, cultural heritage and biodiversity;
- the revolutionary idea that unclaimed mineral resources in the global commons are part of 'the common heritage of mankind', which has been denied much operational relevance in the Law of the Sea treaty, but which might be revived in relation to the equitable and sustainable use of renewable resources such as fresh water, clean air, forests, and aquaculture.

Of course, this recitation of normative ideas invites scepticism. Included in the text of the Rio Declaration are neo-liberal precepts about promoting an open economy and about not burdening trade and investment with environmental restrictions. These assertions represent a regression as compared to the norms sets forth in the Stockholm Declaration. Also, there are glaring discrepancies between the norms affirmed and the behavioural practices of most states, and little political support for proposals to give sustainable development a more operational meaning. Beyond this, there seems to have been little disposition to consider implementing, or even monitoring, mechanisms so as to identify degrees of compliance.

Further, the mood at Rio was resistant to the acceptance of specific commitments even in relation to such urgent global problems as that of climate change, and American leadership, such as it was, served as a depressant in relation to serious initiatives. The dominant mood was neo-liberal in the basic sense of allowing market forces to work efficiently to achieve rapid economic growth accompanied by the expansion of trade and investment, and the sustainable part of sustainable development seemed to count for little. Without commitments to a Tobin tax of some sort to raise funds for environmental governance or to establish an independent global environmental agency, the project to achieve humane governance of the environment must be deemed to remain at a preliminary stage, with most of the hard work lying ahead.

Nevertheless, as with human rights, putting a helpful normative architecture in place is a move towards humane governance. The relevance of this architecture is dependent on implementation from below by transnational social forces, by public pressures generated by the media especially in the setting of environmental disasters, and by unforeseen developments that clip the wings of neo-liberal ascendancy. How to give policy relevance to these norms' bearing on environmental quality provides both challenge and opportunity to global civil society. States have brought the norms into existence, but without serious intention to actualize, and so this energy must come from elsewhere.

Adaptive responses: managerial proposals

In seeking to achieve higher levels of environmental protection without challenging global market forces, several proposals have been put forth. Such proposals assume that 'humane governance' on a global scale can be achieved within, and only within, the basic contours of neo-liberal globalization, that is, without structural or ideological changes.

At the same time, initiatives are needed to avoid the impression that neo-liberal thinking is incapable of or indifferent to environmental problem-solving. In general, the two main types of proposals either involve privatization or volunteerism in some form.

Privatization, in effect, tries to use the Market as a way to diminish environmental threats and harm. An example of privatization would be the marketing of anti-pollution technologies that reduce the task of waste disposal. At Kyoto in 1997, the United States delegation strongly supported the idea of having countries with low emission rates sell 'emission rights' to other countries, as an alternative to imposing strict upper limits on greenhouse gas emission levels. Such a marketization of efforts to address the problem of climate change may be a beneficial way to transfer some resources to poorer countries, but it also undermines the idea of responsible sovereignty with respect to climate change, allowing richer countries to purchase rights to engage in environmentally destructive activity. If, as seems unlikely, it could be shown that even if emission rights were sold, their net impact would not accelerate climate change, then some sort of case could be made in favour of such an initiative.

Volunteerism has become very popular in this period of public sector passivity. In a much-publicized stunt, Ted Turner, the communications magnate, donated a billion dollars to the United Nations over a ten-year period to enable an expanded effort on humanitarian issues, including the environment, and in light of the budgetary pressure on the Organization arising from the failure by important members to pay their dues. The Turner gift signalled the scale of wealth now in the private sector, as well as the sense that voluntary action could fill the vacuum created by governmental inaction. Turner called on other rich individuals to follow in this direction.

In quite another spirit, but still resting on the effectiveness of private sector actions, Kofi Annan, the Secretary General of the United Nations, speaking at the Davos World Economic Forum in early 1999, appealed to business leaders to run their companies in a manner that upheld human rights, labour standards and environmental standards. Annan encouraged such action even in the absence of legal obligations to do so on the part of business. In effect, Annan was contending that upholding these normative obligations should take precedence over the opportunity to gain higher returns on capital by investing in countries with little regulation on these matters. Annan hoped that business leaders would put a sense of responsibility ahead of profits. It seems questionable whether such an appeal would alter behaviour, although perhaps the closely related idea of adopting a code of conduct for overseas investments might at least deter the most abusive arrangements. The failure to abide by such a code could be used by elements of civil society to mount pressure in specific situations, and could be helpful, but hardly capable of achieving comprehensive environmental protection and justice. At most, the gains would be defensive and ad hoc.

Towards humane governance for the environment: a transformative perspective

It seems naive to suppose that environmental protection and fairness can be achieved within the current atmosphere of neo-liberal globalization, in which state actors have accepted a passive role as facilitators. Such a view is strengthened by empirical trends that suggest a worsening of relevant conditions in many respects, both bearing on environmental quality and disparities of use. Such pessimism is reinforced by the prospect of increasing food and water scarcities combined with rapid population growth concentrated in the poorest societies. This overall profile supports the general conclusion that only innovative ideas, values and initiatives of a transformative character have any real prospect of meeting the environmental challenge in a manner that corresponds to humane governance.

There are amid these storm clouds shafts of light that give some reason to be hopeful about the future. To begin with, there is evident a retreat from the unconditional embrace of neo-liberal globalization. The Asian financial crisis, as well as the economic plight of Japan, Russia, Brazil and other countries, have made it clear that the world economy cannot run on automatic pilot. From business leaders there are expressions of concern about 'market fundamentalism', and calls for 'responsible globality', a new financial architecture, a second Bretton Woods, a system of controls on capital flows and currency speculation, a more democratized and open process of policy formation in the World Bank and the IMF. Such reformist measures open space for the consideration of alternatives, including ethical factors that market-driven and realist logic prefers to disregard. Within such space the idea of 'humane global governance' has an almost inevitable place in the search for ways to legitimize modes of governance that are superseding the Westphalian system of states (Falk, 1998). That is, just as the normative claims of territorial states in the late medieval world were based on a blend of claims to competence and values, so the normative claims of an inevitable globalizing world order will need to find practical and idealistic justifications of sufficient appeal to produce a new stability in world order.

Beyond this set of developments, there is visible an interesting series of political moves that suggest a different kind of globalization: what might somewhat simplistically be understood as 'globalization-from-below'. This terminology is designed above all to assert a contrast with capital-oriented 'globalization-from-above', with its nexus of power being the collaboration of large transnational private sector actors and leading states that control the geopolitical agenda. The oppositional forces consist of social movements that are engaged in promoting human rights, environmental protection, developmental equity and substantive democracy, but also new patterns of coalition between 'normal states' and such movements. A 'new international-ism' is evident in global campaigns to ban anti-personnel landmines, to establish an international criminal court, to abolish nuclear weaponry. What gives these campaigns their distinctive political char-acter is the collaboration between a large number of governments and big coalitions of citizens, associations that express grassroots sentiments. Whether this new internationalism can gather further momentum of the sort that might in time begin to fill the power vac-uum created by the weakness of organized labour in a post-industrial global economy is far from assured.

It is also encouraging to take account of the widening and deepening of regional frameworks for cooperation and governance that has been taking place, especially in Europe. It is not only the formation of larger units for competitive participation in the world economy that is impressive, but the weakening of state boundaries, the satisfaction of micro-nationalist identities, the protection of human rights, and the opportunities for democratic participation and accountability. There exists in Europe, despite major problems and disappointments, the possibility of a model of 'compassionate regionalism' entrenching itself in the political imagination of the world, with rapid extension to other important regions including Asia, Africa and Latin America. There is little doubt that European regionalism is already a far more daring and radical experiment in restructuring world order than anything associ-ated with the United Nations. Whether this will eventuate in regional models of human governance cannot be known at this point, but the regional idea seems more promising than any approaches to gover-nance of a global scope.

Of course, deeper adjustments bearing on the rollback of con-sumerism are indispensable if humane governance is to be achieved.

Lifestyles will have to change in fundamental ways if the peoples of the planet are to be fed adequately. Perhaps moves against meat-eating will have to become stronger than the campaigns currently being waged against smoking. In the end, as well, it is almost impossible to entertain the prospect of humane governance without substantial demilitarization of all political relations, and the embrace of non-violent modes of conflict resolution as a matter of fundamental principle. Such changes are beyond the horizon at present, but their relevance to the humane governance of the environment cannot be doubted.

Conclusion

Periods of transition are always murky with respect to trends. Close observers of the international scene missed such cataclysmic changes as the end of the cold war, the collapse of the Soviet Union, the peaceful transition in South Africa. Our knowledge of the forces of change in political life remains primitive. As a result, it is important to remain engaged on behalf of improvements in the overall human condition. The project to establish humane global governance seems far-fetched given dominating ideas and power structures, but there may be concealed fissures in the edifices of authority that create now unknown opportunities for reform and transformation. To take advantage of such fissures, to the extent that they exist, it is important to remain alert to such possibilities. Anything less amounts to a submission to the existing order of governance that seems incapable of meeting the environmental challenge in humane and effective ways.

Bibliography

Abbott, E. *Haiti: The Duvaliers and their Legacy* (London: Robert Hale, 1988).

Abravanel, P. 'L'Environnement et les droits de l'homme', paper presented at the fourth international conference: 'Towards the World Governing of the Environment', Venice, 2–5 June 1994, in Postiglione, A. ed. *Abstracts*, ICEF, 1995, 47.

Altink, S. *Stolen Lives: Trading Women into Sex and Slavery* (London: Scarlet Press, 1995).

Anderson, M. 'Human Rights Approach to Environmental Protection: an Overview', in Boyle, A. *and* Anderson, M. eds. *Human Rights Approaches to Environmental Protection* (Oxford: Oxford University Press, 1996).

Archibugi, D. 'From the United Nations to Cosmopolitan Democracy', in Archibugi, D. and Held, D. eds. *Cosmopolitan Democracy* (Cambridge: Polity, 1995).

Archibugi, D. and Held, D. eds. *Cosmopolitan Democracy. An Agenda for a New World Order* (Cambridge: Polity Press, 1995).

Archibugi, D., and Köhler, M. eds. 'Global Democracy', special issue of *Peace Review*, IX (1998), 309–98.

Archibugi, D., Balduini, S. and Donati, M. 'The United Nations as an Agency of Global Democracy', in Holden, B. ed. *Global Democracy* (London: Routledge, 2000).

Archibugi, D., Held, D. and Köhler, M. eds. *Re-imagining Political Community. Studies in Cosmopolitan Democracy* (Cambridge: Polity Press, 1998).

Atkinson, J. *Undermined: The Impact of Australian Mining Companies in Developing Countries* (Melbourne: Oxfam, 1998).

Bagchi, A. K. 'The GATT Final Act: A Declaration of Rights of TNCs?', *Third World Resurgence*, 46 (1994) 26–9.

Bakhtin, M. *Problems of Dostoevsky's Poetics*, trans. Caryl Emerson (Minneapolis: University of Minnesota Press, 1984).

Barry, B. *Theories of Justice* (London: Harvester Wheatsheaf, 1989).

Baum, J. A. C. and Oliver, C. 'Institutional Linkages and Organizational Mortality', *Administrative Science Quarterly* 36 (1991) 187–218.

Beck, U. 'The Reinvention of Politics: Towards a Theory of Reflexive Modernization', in Beck, U., Giddens, A. and Lash, S. eds. *Reflexive Modernization. Politics, Tradition and Aesthetics in the Modern Social Order* (Cambridge: Polity Press, 1994).

—— 'Risk Society and the Provident State', in Lash, S., Szerszynski, B. and Wynne, B. eds. *Risk, Environment and Modernity* (London: Sage, 1996).

—— 'Democracy Beyond the Nation-State', *Dissent*, XLV (1999) 53–5.

Beckerman, W. *Small is Stupid. Blowing the Whistle on the Greens* (London: Duckworth, 1995).

Beitz, C. *Political Theory and International Relations* (Princeton: Princeton University Press, 1979).

Bellow, S. *Mr. Sammler's Planet* (New York: Viking, 1970).

Bilderbeek, S. 'The Role of the International Court of Justice in Global Environmental Problems', in Postiglione, A. ed. *Tribunale Internazionale per l'Ambiente, Nuovo organo di garanzia dell'ambiente in sede internazionale* (Rome, Istituto Poligrafico e Zecca dello Stato, Libreria dello Stato, 1992).

Birnie, P. W. and Boyle, A. E. *International Law and the Environment* (Oxford: Oxford University Press, 1992).

Black, R. *Refugees, Environment and Development* (New York: Addison-Wesley, Longman, 1998).

Blaug, R. 'Between Fear and Disappointment: Critical, Empirical and Political Uses of Habermas', *Political Studies* 45 (1997) 100–17.

Boehmer-Christiansen, S. 'The International Research Enterprise and Global Environmental Change: Climate-change policy as a research process', in Vogler, J. and Imber, M. F. eds. *The Environment and International Relations* (London and New York: Routledge and ESRC Global Environmental Change Programme, 1996).

Borgman, A. 'Hope, Luck Ran Out on Journey North for Immigrant. Hardships of Road Ended in Smash of Crowded Truck', *Washington Post*, (M) F23-A (1996) 4.

Bosniak, S. L. 'Human Rights, States Sovereignty and the Protection of Undocumented Migrants under the International Migrant Workers Convention', *International Migration Review*, XXV (1994) 737–70.

Bosselmann K. *Im Namen der Natur – Der Weg zum ökologischen Rechtsstaat* (Munich: Scherz, 1992).

—— *When Two Worlds Collide: Society and Ecology* (Auckland: RSVP, 1995).

—— *Ökologische Grundrechte* (Baden-Baden: Nomos, 1998).

Bothe, M. *The Right to a Healthy Environment in the European Union* (London: Kluwer Law International, 1999).

Bourgeois, J. 'A World Authority for the Environment', *Federalist*, 34 (1992) 230.

Boutros-Ghali, B. *An Agenda for Democratization* (New York: United Nations, 1996).

Branigan, W. 'Immigration Fraud Schemes Proliferating Inside US', *Washington Post*, May 19-A, (1996) 1.

—— 'In Garment Workrooms, a Bundle of Abuse and a Thimbleful of Pay', *Washington Post*, F16 (1997) A30.

Brown, M. E., Lynn-Jones, S. M. and Miller, S. E. eds. *Debating the Democratic Peace*. (Cambridge, MA: MIT Press, 1996).

Brown Weiss, E. *In Fairness to Future Generations: International Law, Common Patrimony, and Intergenerational Equity* (Tokyo: Transnational, 1989).

Bruno, K. 'The Corporate Capture of the Earth Summit', *Multinational Monitor* (online), July/August (1992), internet site http://www.essential.org/monitor/hyper/mm0792.html

Cadbury, A. 'The Future of Governance: the Rules of the Game', *Journal of General Management*, 25 (1998) 1–14.

Capra, F. *The Web of Life: A New Scientific Understanding of Living Systems* (New York: Anchor Books, 1997).

Carley, M. and Spapens, P. *Sharing the World. Sustainable Living & Global Equity in the 21st Century* (London: Earthscan Publications, 1998).

Carr, D. *Time, Narrative, and History* (Bloomington: Indiana University Press, 1991).

Carson, R. *Silent Spring* (Cambridge, MA: Riverside Press, 1962).

Casagrande, F. E. 'Sustainable Development in the New Economic [Dis]Order: the Relationship Between Free Trade, Transnational Corporations, International Financial Institutions and Economic Miracles', *Sustainable Development*, 4 (1996) 121–9.

Cela, J. 'Santo Domingo: an Alternative City Plan', in Collinson, H. ed. *Green Guerrillas: Environmental Conflicts and Initiatives in Latin America and the Caribbean*, (London: Latin American Bureau, 1996).

Chossudovsky, M. *The Globalization of Poverty* (London: Zed Books, 1997).

Christiansen, T. 'Space: From Territorial Politics to Multilevel Governance', in Jørgensen, K. E. ed. *Reflective Approaches to European Governance* (London: Macmillan, 1997).

Coalition for Environmentally Responsible Economies (CERES). *The CERES Principles* (amended 28 April 1992), internet site http://www.ceres.org/principles.html

Colchester, M. 'The Struggle for Land: Tribal Peoples in the Face of the Transmigration Programme', *The Ecologist*, 16 (1996) 99–103.

Commission of the European Communities. *Community Strategy and Action Programme for the Forestry Sector* (Brussels, 1988).

Commission on Environment and Development. *Our Common Future* (Oxford: Oxford University Press, 1987).

Commission on Global Governance. *Our Global Neighborhood* (Oxford: Oxford University Press, 1995).

Commission on Sustainable Development. *Introduction to the IPF/IFF Process* (28.1.1999), internet site http://www.un.org/esa/sustdev/iffintro.htm

Connolly, B. 'Increments for the Earth: the Politics of Environmental Aid', in Keohane, R. O. and Levy, M. A. eds. *Institutions for Environmental Aid* (Cambridge, MA and London: MIT Press, 1996).

Cronon, W. *Changes in the Land: Indians, Colonists and the Ecology of New England* (New York: Hill & Wang, 1983).

Crosby, A. *Ecological Imperialism: the Biological Expansion of Europe* (New York: Cambridge University Press, 1986).

Daly, H. E. *Steady State Economics*, 2nd edn (Washington: Island Press, 1991).

—— 'Die Gefahren des freien Handels', *Spektrum der Wissenschaft*, January (1994).

—— Beyond Growth, the Economics of Sustainable Development (Boston: Beacon Press, 1996).

—— 'Sustainable Growth? No Thank You', in Mander, J. and Goldsmith, E. eds. *The Case Against the Global Economy and for a Turn Toward the Local* (San Francisco: Sierra Club Books, 1996).

Daly, H. and Cobb, J. *For the Common Good: Redirecting the Economy toward Community, the Environment, and a Sustainable Future* (Boston: Beacon Press, 1994).

Danquah, A. N. M. 'Life as an Alien', *Washington Post*, May 117-WMAG (1998) 1.

Darier, E. 'Foucault and the Environment: an Introduction', in Darier, E. ed. *Discourses of the Environment* (Oxford: Blackwell Publishers, 1998) 1–34.

Darwish, M. 'Identity card', *New Internationalist*, 277 (1996) 11.

de Campos Mello, V. *North–South Conflicts and Power Distribution in UNCED Negotiations: the Case of Forestry* Working Paper, WP-93-26 (Laxenburg, Austria: International Institute for Applied Systems Analysis, 1993).

Dean, M. *Governmentality* (London: Routledge, 1999).

Deleuze, G. and Guattari, F. *A Thousand Plateaus*, trans. B. Massumi (Minneapolis: University of Minnesota Press, 1988).

Deller, K. and Spangenberg, J. H. *Wie zukunftsfähig ist Deutschland? [How sustainable is Germany?]* (Bonn: Forum Umwelt & Entwicklung, 1997) 25–38.

van Dieren, W. ed. *Taking Nature into Account* (New York: Copernicus Springer-Verlag, 1995).

Diesendorf, M. and Hamilton, C. eds. *Human Ecology, Human Economics: Ideas for an Ecologically Sustainable Future* (St Leonards, Australia: Allen & Unwin, 1997).

Dobson, A. *Justice and the Environment. Conceptions of Environmental Sustainability and Dimensions of Social Justice* (Oxford: Oxford University Press, 1998).

Dryzek, J. 'Global ecological democracy', in Low, N. P. ed. *Global Ethics and Environment* (New York: Routledge, 1999).

Durning, B. A. 'Poverty and the Environment: Reversing the Downward Spiral', *Worldwatch Paper 92* (1989).

Dworkin, R. 'In Defense of Equality', *Social Philosophy and Policy* 1 (1983), 24–40.

Ekins, P. and Max-Neef, M. eds. *Real-Life Economics. Understanding Wealth Creation* (London, New York: Routledge, 1992).

Elias, D. 'Insurers in push for ecology', *The Age*, Melbourne (1999) 16 October, 1, Business Section.

Elliott, A. M. 'Symptoms of Globalization: or, Mapping Reflexivity in the Postmodern Age', in Camilleri, J. A., Jarvis, A. P. and Paolini, A. J. eds. *The State in Transition. Reimagining Political Space* (Boulder: Lynne Rienner, 1995).

Enloe, C. *Making Feminist Sense of International Relations: Bananas, Beaches and Bases* (Berkeley, CA: Pandora, 1990).

European Parliament. *A Global Community Strategy in the Forestry Sector: Summary, Conclusions, Proposals* (Directorate General for Research, Working Papers: Agriculture, Fisheries and Forestry Series, 1994).

European Parliament. *Report on a General Community Strategy for the Forestry Sector – Committee on Agriculture and Rural Development* (Luxembourg: Office for Official Publications of the European Communities, 1996).

Evans, T. 'International Environmental Law and the Challenge of Globalization', in Jewell, T. and Steele, J. eds. *Law in Environmental Decision-Making: National, European and International Perspectives* (Oxford: Oxford University Press, 1998).

Exell Pirro, D. 'Proposal for an International Court of Justice for the Environment within the United Nations', in Postiglione, A. ed. *Tribunale Internazionale per l'Ambiente. Nuovo organo di garanzia dell'ambiente in sede internazionale* (Rome, Istituto Poligrafico e Zecca dello Stato, Libreria dello Stato, 1992) 279–84.

Falk, R. *On Humane Governance: Towards a New Global Politics* (Cambridge: Polity, 1995).

—— *Law in an Emerging Global Village: a Post-Westphalian Perspective* (Ardsley, NY: Transnational Publishers, 1998).

Falkner, R. 'Corporations and Agenda-Setting in International Environmental Politics', paper presented to the ESRC Global Environmental Change meeting, City University, London, 18 April 1997.

Feldman, D. L. 'Iterative Functionalism and Climate Management Organizations: from Intergovernmental Panel on Climate Change to Intergovernmental Negotiating Committee', in Bartlett, R. V., Kurian, P. A. and Malik, M. eds. *International Organizations and Environmental Policy* (London: Greenwood Press, 1995)187–208.

Ferguson, K. *The Man Question: Visions of Subjectivity in Feminist Theory* (Berkeley, CA: University of California Press, 1993).

Fishkin, J. *Democracy and Deliberation* (New Haven: Yale University Press, 1991).

—— *The Voice of the People* (New Haven: Yale University Press, 1997).

Foucault, M. 'The Subject and Power', in Dreyfus, H. and Rabinow, P. *Michel Foucault: Beyond Structuralism and Hermeneutics* (Chicago: University of Chicago Press, 1984) 208–26.

—— *The Use of Pleasure*, trans. Hurley, R. (New York: Pantheon, 1985).

—— 'The Ethics of the Care of the Self as a Practice of Freedom', in Rabinow, P. ed. *Michel Foucault: Ethics, Subjectivity and Truth* (New York: New Press, 1997a) 281–302.

—— 'What is Enlightenment?', in Rabinow, P. ed. *Michel Foucault: Ethics, Subjectivity and Truth* (New York: New Press, 1997b) 303–20.

Fox, M. *A Spirituality Named Compassion* (San Francisco: Harper & Row, 1990).

French, H. *After the Earth Summit: the Future Environmental Governance*, Worldwatch Paper 107 (Washington: Worldwatch Institute, 1992).

Gadgil, M. and Guha, R. *Ecology and Equity* (London & New York: Routledge, 1995).

Gan, L. 'Global Warming and the World Bank: a System in Transition', *Project Appraisal* 8 (1993) 198–212.

Gare, A. *Nihilism Inc.: Environmental Destruction and the Metaphysics of Sustainability* (Sydney: Eco-Logical Press, 1996).

Gemeinsame Erklärung der evangelischen und katholischen Kirche (public statement, 1989).

George, S. *How the Other Half Dies: the Real Reasons for World Hunger*, (Harmondsworth: Penguin, 1976), cited by O'Rourke, P. J. 'All Guns, No Butter', in Sterling, R. ed. *Food: True Stories of Life on the Road* (Sebastopol, CA: O'Reilly & Associates, 1996) 68.

Giagnocavo, C. and Goldstein, H. 'Law Reform or World Re-form', *McGill Law Journal* 35 (1990) 346.

Gibson, N. 'The Right to a Clean Environment', *Saskatchewan Law Review* 54 (1990) 5.

Glacken, C. J. *Traces on the Rhodian Shore* (Berkeley: University of California Press, 1967).

Greene, O. 'Environmental Regimes: Effectiveness and Implementation Review', in Vogler, J. and Imber, M. F. eds. *The Environment and International Relations* (London and New York: Routledge and the ESRC Global Environmental Change Programme, 1996) 196–214.

Habermas, J. *Moral Consciousness and Communicative Action*, Lenhardt, C. and Nicholsen, S. W., trans. (Cambridge, MA: MIT Press, 1995a).

—— 'Reconciliation through the Public Use of Reason', *Journal of Philosophy* XCII, 3 (1995b) 109–31.

Haraway, J. D. *Simians, Cyborgs, and Women* (London and New York: Routledge, 1991).

Hardin, G. *Exploring New Ethics for Survival: the Voyage of the Spaceship Beagle* (New York: Beagle, 1972).

Hathaway, J. ed. *Reconceiving International Refugee Law* (Dordrecht: Kluwer Law International, 1997).

Hayter, T. *Exploited Earth* (London: Earthscan, 1989).

Heater, D. *World Citizenship and Government. The Cosmopolitan Idea in the History of Western Political Thought* (London: Macmillan, 1996).

Heidegger, M. *The Question Concerning Technology and Other Essays,* trans. Lovitt, W. (New York: Harper Torchbooks, 1977).

Held, D. *Democracy and the Global Order: from the Modern State to Cosmopolitan Governance* (Cambridge: Polity, 1995).

—— Models of Democracy, 2nd edn (Cambridge: Polity Press, 1996).

—— 'Democracy and Globalization', *Global Governance,* 3 (1997) 251–65.

—— 'Democracy and Globalization', in Archibugi, D., Held, D. and Kohler, M. eds. *Reimagining Political Community* (Cambridge: Polity Press, 1998) 11–27.

Held, D., McGrew, A., Goldblatt, D. and Perraton, J. *Global Transformations. Politics, Economics and Culture* (Cambridge: Polity Press, 1999).

Hempel, L. *Environmental Governance: the Global Challenge* (Washington, DC: Island Press, 1996).

Henderson, G. M. 'Introduction', in Henderson, G. M. ed. *Borders, Boundaries and Frames: Essays in Cultural Criticism and Cultural Studies* (New York and London: Routledge, 1995).

Hinterberger, F., Messner, D. and Meyer-Stahmer, J. eds. *Complete Paper Series* 1–3 (1997–8).

Hirst, P. and Thompson, G. *Globalization in Question: the International Economy and the Possibility of Governance* (Cambridge: Polity, 1996).

Hobsbawm, E. 'The Death of Neo-Liberalism', *Marxism Today,* November/ December (1998) 4–8.

Hohmann, H. ed. *Basic Documents of International Environmental Law* 3 (1992) 1748.

Holden, B. ed. *Global Democracy* (London: Routledge, 2000).

Homer-Dixon, T. F., Boutwell, J. and Rathjens G. E. 'Environmental Change and Violent Conflict', *Scientific American,* 268 (1993) 38–45.

Horton, R. 'African Traditional Thought and Western Science', *Africa,* 37 (1967) 50–71; 155–87.

Hoyle, F. *The Nature of the Universe* (Oxford: Blackwell, 1950).

Humphreys, D. *Forest Politics. The Evolution of International Cooperation* (London: Earthscan, 1996).

—— 'The Report of the Intergovernmental Panel on Forests', *Environmental Politics,* 7 (1998) 214–21.

Hundertwasser, F. 'The Straight Line', in Peitgen, H. O. and Richter, P. H. eds. *The Beauty of Fractals: Images of Complex Dynamical Systems* (Heidelberg: Springer-Verlag, 1986).

International Chamber of Commerce (ICC). *Business Charter for Sustainable Development* (BCSD) (Paris, April 1991) in *Review of European Community and International Environmental Law,* 4 (1995) 176–7.

International Court of the Environment Foundation. *The History of an Idea,* in Postiglione, A. ed. *Tribunale Internazionale per l'Ambiente. Nuovo organo di garanzia dell'ambiente in sede internazionale* (Rome, Istituto Poligrafico e Zecca dello Stato, Libreria dello Stato, 1992) 67–97.

International Institute for Sustainable Development (IISD) *A Brief Introduction to Global Forest Policy,* internet site: http://www.iisd.ca/linkages/forestry/forestsintro.html (30.10.1998).

Jager, T. and O'Riordan, J. eds. *Politics of Climate Change: A European Perspective,* (London and New York: Routledge, 1996).

Jonkman, P. J. 'Resolution of International Environmental Disputes: a Potential Role for the Permanent Court of Arbitration', paper presented at the Fourth International Conference, *Towards the World Governing of the Environment* (Venice, 2–5 June 1994).

Kaelberer, M. 'Hegemony, Dominance or Leadership? Explaining Germany's Role in European Monetary Cooperation', *European Journal of International Relations,* 3 (1997) 35–60.

Kaldor, M. *New and Old Wars. Organized Violence in a Global Era* (Cambridge: Polity Press, 1998).

Kant, I. 'Perpetual Peace. A Philosophical Sketch' [1795] in Reiss, H. ed. *Political Writings* (Cambridge: Cambridge University Press, 1991).

Keohane, R. O. 'Analyzing the Effectiveness of International Environmental Institutions', in Keohane, R. O. and Levy, M. A. eds. *Institutions for Environmental Aid* (Cambridge, MA and London: MIT Press, 1996) 3–28.

Kimball, L. A. *Forging International Agreement* (Washington: World Resources Institute, 1992).

Kirchenamt der Evangelischen Kirche in Deutschland, Sekretariat der Deutschen Bischofskonferenz *Für eine Zukunft in Solidarität und Gerechtigkeit* [*Towards a Future in Solidarity and Justice*]. *Wort des Rates der Evangelischen Kirche in Deutschland und der deutschen Bischofskonferenz zur wirtschaftlichen und sozialen Lage* (Hannover/Bonn: Kirchenamt Sekretariat, 1997).

Kiss, A. and Shelton, D. *International Environmental Law* (Dordrecht: Transnational, 1991).

Knudtson, P. and Suzuki, D. *Wisdom of the Elders* (Toronto: Stoddard Press, 1992).

Kolk, A. *Forests in International Environmental Politics* (Utrecht: International Books, 1996).

Korten, D. C. *When Corporations Rule the World* (London: Earthscan, 1995).

Kux, S. 'Pushers, laggards and free-riders: Explaining the international framework and the domestic bases of climate politics', paper prepared for the Small Group Session 'Environmental Governance' in *1997 Open Meeting of the Human Dimensions of Global Change Research Community* (IIASA, Laxenburg, Austria, 12–14 June 1997).

Kymlicka, W. *Multicultural Citizenship* (Oxford: Oxford University Press, 1995).

Laden, A. *Constructing Shared Wills: Deliberative Liberalism and the Politics of Identity,* unpublished PhD dissertation, Department of Philosophy, Harvard University, 1997.

Lasch, C. *The Revolt of the Elites and the Betrayal of Democracy* (New York: W.W. Norton, 1995).

Laslett, P. '*Is there a generational contract?*', in Laslett, P. and Fishkin, J. *Justice between Age Groups and Generations* (New Haven: Yale University Press, 1992).

——— *The World We Have Lost: Further Explored,* 3rd edn (London: Methuen, [1965]1983).

Laslett, P. and Fishkin, J. *Justice between Age Groups and Generations* (New Haven: Yale University Press, 1992).

Lee, K., Humphreys, D. and Pugh, M. '"Privatisation" in the United Nations System: Patterns of Influence in Three Intergovernmental Organisations', *Global Society: Journal of Interdisciplinary International Relations* 11 (1997) 339–57.

Levy, M. A., Keohane, R. O. and Haas, P. M. 'Improving the Effectiveness of International Environmental Institutions', in Haas, P. M., Keohane, R. O. and Levy, M. A. eds. *Institutions for the Earth: Sources of Effective International Environmental Protection* (Cambridge MA and London: MIT Press, 1993) 397–426.

Lewis, S. 'Stakeholder Audits: an Alternative Approach to Environmental Auditing', *Review of European Community and International Environmental Law* 4 (1995) 123–32.

Liberatore, A. 'The European Union: bridging domestic and international environmental policy-making', in Scheurs, M. A. and Economy, E. C. eds. *The Internationalization of Environmental Protection* (Cambridge: Cambridge University Press, 1997).

Linklater, A. *The Transformation of Political Community* (Cambridge: Polity Press, 1998).

Lorek, S. and Spangenberg, J. H. *Priorities and Indicators of Environmentally Sound Household Consumption* (Witten, Dortmund: pad-Verlag, 1999).

Low, N. ed. *Global Ethics and Environment* (London: Routledge, 1999).

Low, N. and Gleeson, B. *Justice, Society & Nature: an Exploration of Political Ecology* (London: Routledge, 1998).

Luper-Foy, S. 'International justice and the environment', in Cooper, D. E. and Palmer, J. A. eds. *Just Environments* (London: Routledge, 1995).

Lyotard, J. *The Postmodern Condition: A Report on Knowledge*, trans. Bennington, G. and Massumi, B. (Minneapolis: University of Minnesota Press, 1984).

MacIntyre, A. 'Epistemological Crises, Dramatic Narratives and the Philosophy of Science', *Monist*, 60 (1977) 453–72.

—— *After Virtue*, 2nd edn (Notre Dame, IND: University of Notre Dame Press, 1984).

Mander, J. and Goldsmith, E. eds. *The Case Against the Global Economy and For a Turn Toward the Local* (San Francisco: Sierra Club Books, 1996).

Marchak, M. P. *Logging the Globe* (Montreal: McGill Queens University Press, 1995).

Marchand, G. 'Towards the Overcoming of Absolute National Sovereignty', *The Federalist*, 34 (1992) 233.

Marks, G., Scharpf, F. W., Schmitter, P. C. and Streeck, W. 'Preface', in Marks, G., Scharpf, F. W., Schmitter, P. C. and Streeck, W. eds. *Governance in the European Union* (London: Sage, 1996).

Mathews, F. *The Ecological Self* (London: Routledge, 1991).

McCully, P. *Silenced Rivers: the Ecology and Politics of Large Rivers* (London and New Jersey: Zed Books, 1996).

McDougal, M. S., Lasswell, H. D. and Chen, L. *Human Rights and World Public Order* (New Haven: Yale University Press, 1980).

McDrew, A. *The Transformation of Democracy?* (Cambridge: Polity Press, 1997).

Meadows, D. H., Meadows, D. L., Randers, J. and Behrens, W. W. *The Limits to Growth* (New York, Universe Books, 1972).

Merchant, C. *Ecological Revolutions: Nature, Gender, and Science in New England* (London: University of North Carolina Press, 1989).

Merchant, C. *Radical Ecology: the Search for a Livable World* (London: Routledge, 1992).

Messner, D. *Die Netzwerkgesellschaft. Wirtschaftliche Entwicklung und internationale Wettbewebsfähigkeit als Probleme gesellschaftlicher Steuerung* [*The Network Society. Economic development and international competitiveness as problems of governance*] (Cologne: Weltforum-Verlag, 1995).

M'Gonigle, M. *Forestopia* (Vancouver: Harbour Publishing, 1994).

Midgley, M. *Evolution as a Religion* (London: Methuen, 1985).

Ministerial Conference on the Protection of Forests in Europe. *Interim Report on the Follow-up of the Second Ministerial Conference* (Helsinki: Ministry of Agriculture and Forestry, 1995).

Mintzer, I. M. and Leonard, J. A. eds. *Negotiating Climate Change: the Inside Story of the Rio Convention* (Cambridge: Cambridge University Press and Stockholm Environment Institute, 1994).

Mol, A. P. J. 'Ecological Modernization and Institutional Reflexivity: Environmental Reform in the Late Modern Age', *Environmental Politics*, 5 (1996) 302–23.

Myers, N. 'The Environment Dimension to Security Issues', *Environmentalist*, 6 (1986) 251–7.

Myers, N. *Environmental Exodus: an Emergent Crisis in the Global Arena* (Washington, DC: Climate Institute, 1995).

Nader, R. and Wallach, L. 'GATT, NAFTA and the Subversion of the Democratic Process', in Mander, J. and Goldsmith, E. eds. *The Case Against the Global Economy and For a Turn Toward the Local* (San Francisco: Sierra Club Books, 1996).

NGO *Alternative Treaties*. 'Treaty on Transnational Corporations: Democratic Regulation of their Conduct', internet site http://www.igc.apc.org/habitat/treaties (Global Forum, Rio de Janeiro, 1–15 June 1992).

Nickel, J. W. 'The Human Right to a Safe Environment: Philosophical Perspectives on Its Scope and Justification', *Yale Journal of International Law*, 18 (1993) 281.

Nussbaum, Martha, 'Aristotelian Social Democracy', in Douglass, R., Mara, G. and Richardson, H. eds, *Liberalism and the Good* (New York and London: Routledge, 1990) 207–8.

O'Connor, J. *The Meaning of Crisis: a Theoretical Introduction* (Oxford: Blackwell, 1987).

Ökologische Briefe. 'Die soziale Dimension des Konsums' (the social dimension of consumption) Ökologische Briefe 51/52 (1995) 7 [cited in von Weizsäcker, E. U., Lovins, A. B. and Lovins, H. *Faktor*, 4 (Munich: Droemer Knaur, 1995)].

Ophuls, W. *Ecology and the Politics of Scarcity Revisited: the Unravelling of the American Dream* (New York: W. H. Freeman 1992).

Organisation for Economic Co-operation and Development (OECD). 'Responsibility and Liability of States in Relation to Transfrontier Pollution', *Environmental Policy and Law*, 13 (1984) 122.

Organisation for Economic Co-operation and Development (OECD). 'Thinking about taxation' (Cover Issue) OECD Observer, 215 (1999) 3–23.

O'Rourke, P. J. 'All guns, no butter', in Sterling, R. ed. *Food: True Stories of Life on the Road* (Sebastopol, CA: O'Reilly and Associates, 1996).

Palmer, G. 'New Ways to Make Environmental Law', *American Journal of International Law*, 86 (1992) 259–83.

Paterson, M. *Global Warming and Global Politics* (London and New York: Routledge, 1996).

Pearce, F. *Green Warriors: the People and the Politics behind Environmental Revolution* (Oxford: Bodley Head, 1991).

Pernthaler, P. 'Reform der Bundesverfassung im Sinne des ökologischen Prinzips', in Pernthale, P., Weber, K. and Wimmer, N. eds. *Umweltpolitik durch Recht* (Vienna: Manz, 1992).

Peterson, S. V. 'Transgressing Boundaries: Theories of Knowledge, Gender and International Relations', *Millennium*, 21 (1992) 183–206.

Pettman, J. J. *Wordling Women: a Feminist International Politics* (Sydney: Allen & Unwin, 1996).

van der Pijl, K. 'The Second Glorious Revolution: globalising elites and historical change', in Hettne, B. ed. *International Political Economy, Understanding Global Disorder* (London and New Jersey: Zed Books, 1995).

Plumwood, V. *Feminism and the Mastery of Nature* (London: Routledge, 1993).

Pogge, T. W. 'An Egalitarian Law of People', *Philosophy and Public Affairs*, 23 (1994) 195–224.

—— 'Eine globale Rohstoffdividende', *Analyse und Kritik*, 17 (1995) 183–208.

Porter, G. and Brown, J. W. *Global Environmental Politics*, 2nd edn (Boulder: Westview Press, 1996).

Postiglione, A. *The Global Village without Regulation* (Florence, 1994).

—— *Towards the World Governing of the Environment* (Rome: International Court of the Environment Foundation, 1995).

Pritchett-Post, E. S. *Women in Modern Albania: First Hand Accounts of Culture and Conditions from over 200 Interviews* (Jefferson, NC: Mcfarland, 1998).

Preuss, Ulrich K., 'Citizenship and Identity: Aspects of a Political Theory of Citizenship', in Bellamy, R., Buffacchi, V. and Castiglione, D. eds, *Democracy and the Constitutional Culture in the Union of Europe* (London: Lothian Foundation Press, 1995) 117.

Purdue, D. 'Hegemonic Trips: World Trade, Intellectual Property and Biodiversity', *Environmental Politics*, 4 (1995) 88–107.

Rawls, J. 'Reply to Habermas', *Journal of Philosophy* XCII, 3 (1995) 132–78.

—— *Political Liberalism* (New York: Columbia University Press, 1993).

—— *A Theory of Justice* (Cambridge, MA: Harvard University Press, 1971).

Redgwell, C. 'A Critique of Anthropocentric Rights', in Boyle, A. and Anderson, M. eds. *Human Rights Approaches to Environmental Protection* (Oxford: Oxford University Press, 1996).

Regan, T. 'Does Environmental Ethics Rest on a Mistake?', *Monist*, 75 (1992) 161–82.

Rehg, W. *Insight and Solidarity: a Study of the Discourse Ethics of Jurgen Habermas* (Cambridge, MA: MIT Press, 1994).

Renner, A. ' "Zukunftsfähiges Deutschland" und Ordoliberalismus der Freiburger Schule – zwei gegensätzliche Welten?' [Sustainable Germany and the Ordoliberalism] in Renner, A. and Hinterberger, F. eds. *Zukunftsfähigkeit und Neoliberalismus* (Baden-Baden: Nomos, 1998).

Rich, B. *Mortgaging the Earth* (Boston: Beacon Press, 1994).

Richmond, H. A. *Global Apartheid, Refugees, Racism and the New World Order* (Toronto and Oxford: Oxford University Press, 1994).

Ricoeur, P. *Time and Narrative*, vol. 1, trans. McLaughlin, K. and Pellauer, D. (Chicago: University of Chicago Press, 1984).

—— *Lectures on Ideology and Utopia* (New York: Columbia University Press, 1986).

Rosenau, J. *Along the Domestic–Foreign Frontier. Exploring Governance in a Turbulent World* (Cambridge, Cambridge University Press, 1997).

Rosenau, J. N. and Czempiel, E. O. eds. *Governance without Government: Order and Change in World Politics* (Cambridge: Cambridge University Press, 1992).

Rowell, A. *Green Backlash: Global Subversion of the Environmental Movement* (London: Routledge, 1996).

Russett, B. *Grasping the Democratic Peace* (Princeton: Princeton University Press, 1993).

Rutherford, P. 'The Entry of Life into History', in Darier, E. ed. *Discourses of the Environment* (Oxford: Blackwell, 1998).

Sachs, W., Loske, R. and Linz, M. *Greening the North. A Post-Industrial Blueprint for Ecology and Equity* (London: Zed Books, 1998).

Saladin, P. *Wozu noch Staaten?* (Berne: Stämpfli, 1995).

Sale, K. *Human Scale* (New York: Coward, Cann & Geoghegan, 1980).

—— 'Principles of Bioregionalism', in Mander, J. and Goldsmith, E. eds. *The Case Against the Global Economy and For a Turn Toward the Local* (San Francisco: Sierra Club Books, 1996).

Salleh, A. *Ecofeminism as Politics: Nature, Marx and the Postmodern* (London and New York: Zed Books, 1997).

Sand, P. H. 'New Approaches to Transnational Environmental Disputes', *International Environmental Affairs*, 3 (1991) 193–206.

Sandel, M. *Liberalism and the Limits of Justice* (Cambridge: Cambridge University Press, 1982).

Schatzki, T. *Social Practices: a Wittgensteinian Approach to Human Activity and the Social* (Cambridge: Cambridge University Press, 1996).

Scherhorn, G. *Kaufsucht. Bericht über eine empirische Untersuchung. [Addictive buying. Report on an empirical analysis]* (Stuttgart: Hohenheim University, Institute for Household and Consumption Economics, 1991).

Scherhorn, G. 'Über Konsumentenverhalten und Wertewandel – Die Notwendigkeit der Selbstbestimmung [Consumer behaviour and value changes – the need for self-determination]', *Politische Ökologie*, 34 (1993) 17–23.

Schmidt-Bleek, F. 'Eco-Restructuring Economies: Operationalising the Sustainability Concept', *Fresenius Environmental Bulletin*, 1 (1992) 40–5.

—— 'Towards universal ecological disturbance measures', *Regulatory Toxicology and Pharmacology*, 18 (1993) 456–62.

—— *Carnoules Declaration of the Factor Ten Club* (Wuppertal: Wuppertal Institut für Klima, Umwelt, Energie, 1994).

Schwartz, M. L. 'International Legal Protection for Victims of Environmental Abuse', *Yale Journal of International Law*, 18 (1993) 355.

Seabrook, J. *Victims of Development* (London: Verso, 1993).

Sebenius, J. K. 'Towards a Winning Climate Coalition', in Mintzer, I. M. and Leonard, J. A. eds. *Negotiating Climate Change: the Inside Story of the Rio Convention* (Cambridge: Cambridge University Press and Stockholm Environment Institute, 1994).

Second Session of the Intergovernmental Forum on Forests *Earth Negotiations Bulletin*, internet site: http://www.iisd.ca/linkages/download/asc/enb1345e.txt (25.1.1999).

Shelton, D. 'Human Rights, Environmental Rights, and the Right to Environment', *Stanford Journal of International Law*, 28 (1991) 103.

Shue, H. 'The Unavoidability of Justice', in Hurrell, A. and Kingsbury, B. eds. *The International Politics of the Environment* (Oxford: Clarendon Press, 1992).

Sieghart, P. *The Lawful Rights of Mankind: an Introduction to the International Legal Code of Human Rights* (Oxford: Oxford University Press, 1985).

Singh, N. 'The Environmental Law of War and the Future of Mankind', in Dupuy, R. ed. *The Future of the International Law of the Environment* (The Hague: Martinus Nijhoff, 1985).

Skelton, D. 'Human Rights and Environmental Rights, and the Right to Environment', *Stanford Journal of International Law*, 28 (1991) 103–38.

Smith, A. 'The Law of Nations' [1766] in Meek, R. L., Raphael, D. D. and Stein, P. G. eds. *Lecture on Jurisprudence* (Oxford: Oxford University Press, 1978).

Smith, J. A. 'The CERES Principles: Model for Public Environmental Accountability', *Review of European Community and International Environmental Law*, 4 (1995) 116–22.

Smouts, M. C. 'Multilateralism from Below: a Prerequisite for Global Governance', in Schechter, M. G. ed. *Future Multilateralism: the Political and Social Framework* (Tokyo: United Nations University Press, 1998).

Soroos, M. *The Endangered Atmosphere: Preserving the Global Commons* (Columbia: University of South Carolina Press, 1997).

Soroos, M. S. and Nikitina, E. N. 'The World Meteorological Organization as a Purveyor of Global Public Goods', in Bartlett, R. V., Kurian, P. A. and Malik, M. eds. *International Organizations and Environmental Policy* (London: Greenwood Press, 1995).

Spangenberg, J. H. *Towards Sustainable Europe, a Study from the Wuppertal Institute for Friends of the Earth Europe* (Luton: Friends of the Earth, 1995).

Spangenberg, J. H., Femia, A., Hinterberger, F., Schütz, H. and Moll, S. *Material Flow Based Indicators for Environmental Reporting* (Copenhagen: European Environment Agency, Expert Corner Series, 1999).

Steinberg, R. *Der ökologische Verfassungsstaat* (Frankfurt: Suhrkamp, 1998).

Steward, F. 'Citizens of Planet Earth', in Andrews, G. ed. *Citizenship* (London: Lawrence & Wishart, 1991).

Stone, C. 'Should Tress Have Standing?', *Southern California Law Review* 45 (1972) 1.

Stone, C. D. 'Defending the Global Commons', in Sands, P. ed. *Greening International Law* (London: Earthscan, 1993).

Suggett, D. and Baker, G. 'Kemcor Australia and the Principles of Sustainability', *Review of European Community and International Environmental Law*, 4 (1995) 168–71.

Swanson, B. 'History, Culture and Nature', *The Trumpeter: Journal of Ecosophy*, 8 (1991) 164–9.

Sylvester, C. *Feminist Theory and International Relations in a Postmodern Era* (Cambridge: Cambridge University Press, 1994).

—— 'African and Western Feminisms: World Travelling, the Tendencies and Possibilities', *Signs: Journal of Women in Culture and Development* (1995) 941–69.

—— *Feminist Arts and International Relations* (Copenhagen: Danish Institute of International Affairs, Working Paper 1997/98).

Taylor, C. 'Reply and Re-Articulation', in Tully, J. ed. *Philosophy in an Age of Pluralism* (Cambridge: Cambridge University Press, 1994).

Taylor, P. E. 'From Environmental to Ecological Human Rights: a New Dynamic in International Law?', *Georgetown International Environmental Law Review*, 10 (1998) 309.

Taylor, P. W. *Respect for Nature: an Introduction to Environmental Ethics* (Princeton, NJ: Princeton University Press, 1986).

Temple Lang, J. 'Biological Conservation and Biological Diversity', in Sjöstedt, G. ed. *International Environmental Negotiation* (Newbury Park: IIASA and Sage, 1993).

Third World Resurgence. 'The "Sustainable Council for Business Development"', 24/25 (1992) 22.

Thomas, L. M. 'The Business Charter for Sustainable Development: Action Beyond UNCED', *Review of European Community and International Environmental Law*, 1 (1992) 325–7.

Thompson, J. 'A Refutation of Environmental Ethics', *Environmental Ethics*, 12 (1990) 147–60.

Timoshenko, A. S. 'Ecological Security: Response to Global Challenges', in Brown-Weiss, E. ed. *Environmental Change and International Law: New Challenges and Dimensions* (Tokyo: United Nations University Press, 1992).

Tuit, P. *False Images: Laws Construction of the Refugee* (London: Pluto Press, 1996).

Tully, J. *Strange Multiplicity: Constitutionalism in an Age of Diversity* (Cambridge: Cambridge University Press, 1995).

—— 'To Think and Act Differently: Foucault's Four Objections to Habermas', in Ashenden, S. and Owen, D. eds. *Foucault contra Habermas* (London: Sage, 1999).

United Nations. *Human Development Report* (New York: UN, 1999).

United Nations Centre on Transnational Corporations (UNCTC) 'Code of Conduct for Transnational Corporations', draft (New York, UN, 1992).

United Nations Conference on Environment and Development (UNCED) *Non-legally binding authoritative statement of principles for a global consensus on the management, conservation and sustainable development of all types of forests*, Rio de Janeiro, June 1992, sourced at: gopher://gopher.un.org:70/00/conf/unced/English/forestp.txt (28.1.1999).

United Nations High Commissioner for Refugees (UNHCR) 'Trafficking in Human Lives', *Refugees* (Focus: Asylum in Europe), 101 (1995) 111.

—— *The State of the World's Refugees* (Geneva: UNHCR, 1995).

v.d. Pforten, D. *Ökologische Ethik* (Hamburg: Rowohlt, 1996).

Vidal, J. *McLibel: Burger Culture on Trial*, (London: Macmillan, 1997a).

—— 'The corporate planet', *BBC Wildlife*, July (1997b) 30–1.

Waks, L. J. 'Environmental Claims and Citizen Rights', *Environmental Ethics* 18 (1996) 133–48.

Warren, J. K. ed. *Ecological Feminism* (London: Routledge, 1994).

Watson, H. with the Yolngu community at Yirrkala, *Singing the Land, Signing the Land* (Australia: Deakin University Press, 1989).

Weinberg, B. *War on the Land: Ecology and Politics in Central America* (London: Zed Books, 1991).

von Weizsäcker, C. 'Error-friendliness and the deliberate release of GMOs', in Leskien, D. et al. eds. *European Workshop on Law and Biotechnology, Proceedings* (Bonn: BBU-Verlag, 1990) 42–7.

von Weizsäcker, E. U., Lovins, A. B., Lovins, H. *Faktor* 4 (Munich: Droemer Knaur, 1995).

Welford, R. 'Business and environmental policies', in Blowers, A. and Glasbergen, P. eds. *Environmental Policy in an International Context: Prospects for Environmental Change* (London: Arnold, 1996).

Weston, B. H. 'Human Rights', *Human Rights Quarterly,* 6 (1986) 257.

Weston, B. H., Falk, R. and Charlesworth, H. eds. *Supplement of Basic Documents to International Law and World Order,* 3rd edn (St Paul, MN: West, 1997).

Weterings, R. and Opschoor, J. B. 'Environmental utilization space and reference values for performance evaluation', *Milieu,* 9 (1994) 221–8.

Wiener, J. 'Money Laundering: Transnational Criminals, Globalisation and the Forces of "Redomestication"', *Journal of Money Laundering Control,* 1 (1997) 51–63.

Wiener, J. and Kennair, J. 'Foreign Direct Investment: the Globalisation of Ideology and Law and the Regionalisation of Politics', in Chan, S. and Wiener, J. eds. *Twentieth Century International History: a Reader* (London: I. B. Tauris, 1998).

Wilson, J. *Talk and Log: Wilderness Politics in British Columbia* (Vancouver: University of British Columbia Press, 1998).

Wind, M. 'Rediscovering Institutions: a Reflectivist Critique of Rational Institutionalism', in Jørgensen, K. E. ed. *Reflective Approaches to European Governance* (London: Macmillan, 1997).

Wiyaprao, P. 'Interview', in Ekachai, S. 'Finding jobs for the girls', *Bangkok Post,* 6 August (1996).

World Commission on Environment and Development (WCED) *Our Common Future* (Oxford and New York: Oxford University Press, 1987).

World Resources Institute. *Recent History of International Inter-governmental Forest Policy-Making,* internet site: http://www.wri.org/biodiv/opp-iii.html (25.1.1999).

World Wide Fund for Nature (WWF) UNCED: *the Way Forward* (Gland, Switzerland: WWF-International, 1992).

Yablokov, A., Zabelin, S., Iemeshu, M., Rervina, S., Flerova, G. and Cherkazova, M. 'Russia Gasping for Breath, Choking in Waste, Dying Young', *Washington Post* 18 August (1993) C3.

York, W. *Transnational Corporations in World Development* (New York: United Nations Centre on Transnational Corporations, 1988).

Young, I. M. 'Communication and the Other: Beyond Deliberative Democracy', in Benhabib, S. ed. *Democracy and Difference* (Princeton: Princeton University Press, 1996) 120–36.

Young, O. R. *International Governance: Protecting the Environment in a Stateless Society* (London: Cornell University Press, 1994).

Yunupingu, G. ed. *Our Land is our Life* (St Lucia: University of Queensland Press, 1997).

Index

Page numbers in bold type, e.g. **29–43**, refer to detailed discussion of a topic.

anarchism, 13, 166–7, 178
androcentrism *see* women
anthropocentrism, 10–11, 24, 121,
   124–6, 129–30
arbitration, 18–19, **216–18**
autonomy, 12, 16, 148, 192, 194

Biodiversity Convention, 130–1
Business Charter for Sustainable
   Development (BCSD), 8, **92–3**, 99

capitalism, 4, 13, 40, 163, 167
citizen's juries, 14, 175
citizenship *see* planetary citizenship
climate change management, 2, 5–6,
   44, 91
   see also Intergovernmental Panel on
     Climate Change (IPCC)
Coalition for Environmentally
   Responsible Economies (CERES),
   8, **93–4**, 99
codes of behaviour
   environmental refugees, 85
   TNCs, 2, 8, **92–4**
   UN, 8, 88–93, **94**, 99
conflict resolution, 12, 147–9, 156
constitutions, 10, 23, **122–4**, 128
consumption, 19, 37–9
control, 9, 105, 153–4
cooperation
   international environmental
     management, 54
   planetary citizenship, 11–12, 135–6,
     142–4
cosmologies, 112–13
cosmopolitan democracy, 11–12,
   135–8
   conflict resolution, 147–8
   environmental justice, 16, **192–5**
   governance, 144–6
   international law, 8

planetary citizenship, 139–41
   TNCs, 89, 97–9
cosmopolitical democracy, 16–18, 21,
   196–210
cultural diversity, 2, 11, 135
   ecological ethics, 12, 147–9

deliberative democracy, 14, 16, 22,
   175–7
dematerialization, 4, **31–2**, 35
democracy, 3–4, 7, 40
   legitimacy, 2, 148, 151
   process, 14, 173–6
   see also cosmopolitan democracy;
     cosmopolitical democracy;
     deliberative democracy; global
     democracy
deterritorialization *see* environmental
   refugees
development *see* environmental
   refugees; 'North–South'
   differences
discourse *see* grand narratives
displacement *see* environmental
   refugees
disputes *see* international
   environmental disputes
distributional justice, 1, 23
   sustainability, 3–4, **29–31**, 35

earth citizenship *see* planetary
   citizenship
Earth Summits *see* United National
   Conference on Environment and
   Development
ecological ethics, 9–14, 21–2, **147–64**
   human rights, 124–33
   implementation and review, 162–3
   negotiations over central question,
     158–62
   practical systems, 153–7

ecological rights *see* environmental
   rights
ecological governance *see*
   environmental governance
economics
   anthropocentrism, 11
   democratization, 16
   discourse, 9
   disembedded, 7
   forest conservation, 6
   sustainability, 1, 3–4, **31–2**, 38–41
   see also globalization; progress;
      transnational corporations
emissions trading, 5–6, 50–1, 55
environmental disputes *see*
   international environmental
   disputes
environmental ethics *see* ecological
   ethics
environmental governance, 2–3, **21–5**
   climate change, 6, 57
   TNCs, 94–9
environmental justice
   democracy, 189–95
   deterritorialization, 7
   liberal-egalitarian conception,
      15–16, **183–9**
   political institutions, 13–14, 165–6,
      169
   refugees, 72–3, **84–6**
environmental refugees, 2, 7, **72–87**
   definition, 74–5
   justice maximization, 72–3, **84–6**
environmental rights, 10, **119–24**,
   148
   planetary citizenship, 141–2
epistemic communities *see* scientific
   networks
ethical governance *see* environmental
   governance
ethics, 2–3
   dilemmas, 3–9
   ecological *see* ecological ethics
   global *see* global ethics
European Round Table of
   Industrialists, 91
European Union (EU)
   environmental human rights, 122–4
   forest policy, 6–7, 62, **66–71**

regionalism, 235

failures, as learning, 4–5
Food and Agriculture Organization
   (FAO), 63
forest conservation, 2, 6–7
   European Union, 6–7, 62, **66–71**
   international politics, 6–7, **61–71**
   UNCED policy, 63–4
Framework Convention on Climate
   Change (FCCC), 5, 50–1, 55, 57,
   59

General Agreement on Trade and
   Tariffs (GATT), 91
Global Atmospheric Watch (GAW), 50
Global Climate Coalition, 91
Global Climate Observing System
   (GCOS), 50
global democracy, 15–16, **183–95**
Global Environment Facility (GEF),
   50–1
global ethics *see* ecological ethics
global governance, 2, 5, 8–11, 15–16,
   22, 25, 96, 105–6, 116, 183, 193
global government, 6, 9, 22–3, 25
Global Ocean Observing System
   (GOOS), 50
Global Resource Dividend (GRD),
   15–16, **185–7**, 189, 191, 194
Global Terrestrial Observing System
   (GTOS), 50
global warming *see* climate change
   management
globalization
   environmental refugees, 74
   governance, 1–2, 22, 25
   grand narratives, 111
   humane governance, 223–4, 234
   nation states, 2, 8, **199–200**
   TNCs, 89, 100
good, 108, 110–11, 113
governance, 6
   environmental refugees, 72–3, 84–6
   forest conservation, 61
   see also cosmopolitan governance;
      environmental governance;
      global *see* global governance;
      humane governance

government *see* global government
grand narratives, 2, 9–10, **105–17**
  polyphonic, 2, 10, 114–17

Human Dimensions Programme
  (HDP), 50
human rights, 3, 10, 40, **118–34**
  cosmopolitanism, 135, 137
  environmental refugees, 7, 72–3
  international, 119–22
  see also humanitarian interventions
human values *see* ethics
humane governance, 2–3, 14, 19–21,
  221–36
humanitarian interventions, 17–18,
  21, **206–10**

institutions, 9, 22–3, 39–40, 106, 112,
  162–3
  see also international institutions;
    political institutions
Intellectual Property Coalition, 91
Intergovernmental Forum on Forests
  (IFF), 6, 66
Intergovernmental Panel on Climate
  Change (IPCC), 5–6, **44–60**
  agreement negotiation, 5–6, 44, 48,
    50, **51–5**, 57–9
  compliance and verification, 5–6,
    44–5, **55–8**
  context of activity, 50–1
  Data Distribution Centre (DDC), 57
  epistemic communities, 52–3, 56,
    59
  membership and organizational
    structure, 46–8
  working groups, 5, 46–7, **48–9**,
    50–4, 57
Intergovernmental Panel on Forests
  (IPF), 6, **64–5**
International Chamber of Commerce
  (ICC), 91–2
International Court of Justice, 18,
  211, 214–16, 219
International Court of the
  Environment (proposed), 8–9,
  18–19, 22, 98, **211–20**
international environmental disputes,
  18–19, 24, **211–13**

arbitration, 18–19, **216–18**
jurisdiction, 213–16
International Geosphere Biosphere
  Programme (IGBP), 50
international institutions, 2, 5–6, 8,
  11
  climate change management, 57,
    59
  cosmopolitanism, 135, 137
  democracy, 201, 205
international law
  climate change management, 45,
    59
  cosmopolitan democracy, 8
  human rights, 119
  TNCs, 95, 97–9
  see also International Court of the
    Environment
International Monetary Fund (IMF) /
  World Bank, 32, 234
  environmental refugees, 79–80
international politics
  forest conservation, 6–7, 61–71
  procedural leadership and brokers,
    70–1
  sustainability, 3–4, **39–42**
intervention *see* humanitarian
  interventions

jurisdiction, 213–16
justice, 108, 110–11, 113, 115–16
  cosmopolitanism, 135–7
  ecological ethics, 12–13, 147–50
  see also distributional justice;
    environmental justice

labour, 4, **34–7**
lifestyles, 37–9

market forces, 3–4, 40, 112
  humane governance, 19–20, 223
  TNCs, 89–91
migration *see* environmental refugees
modernity, 9–11, 105–6, 111, 115–16
  Europe, 66–8
Multinational Agreement on
  Investment (MAI), 100
multinational corporations *see*
  transnational corporations

narratives *see* grand narratives
nation states
  climate change management, 5, 54,
    56
  cooperation, 11, 136–8, 144
  cosmopolitical democracy, 16–17
  ecological human rights, 10
  forest conservation, 6–7
  globalization, 2, 8, 22, 25, **199–200**
  humane governance, 20–1, 222–3,
    227
  international system, 200–3
  obsolescence, 13–14, 166–8, 171–2,
    174–5
  political authority, 16, **196–8**
  sustainability, 4, 8
  transnational corporations, 13,
    89–92, 94–8, 100
  Westphalian sovereignty, 17, 19, 67,
    234
nature, 10, 119
  ecological ethics, 12–13, 147–9
  environmental justice, 13, 165, 169,
    178
neo-liberalism, 7, 9, 19–21, 89,
  221–36
non-governmental organizations
  (NGOs)
  climate change management, 53
  political representation, 17
  TNC regulation, 90, 93, 99
'North–South' differences, 3
  cosmopolitan democracy, 97
  environmental justice, 23–4, 220
  forest conservation, 6, 64–5
  humane governance, 229–30
  sustainability, 36

Organisation for Economic Cooperation and Development
  (OECD)
  human rights, 121
  Multinational Agreement on
    Investment (MAI), 100

planetary citizenship, 11, 21, **135–46**
  cooperation, 142–4
  transnational elite, 92
political institutions

nationalism, 205–6
  obsolescence, 13–14, **165–79**
  political parties and process, 17, 22
  TNCs, 8, 90–1, 95, 97
politics *see* international politics
population displacement *see*
  environmental refugees
postmodernist philosophy, 9, 11, 112,
  115
poverty, 15–16, **185–9**, 194–5
privatization, 20, 89–90, 100, 233
production, 4, 23, 39
progress, 106, 112
public charter, 8, 95

redomestication, 8, 95
refugees *see* environmental refugees
religion *see* spirituality
representative democracy, 12, 22,
  151–2
resettlement *see* environmental
  refugees
resource equity principle, 15, 185
resource productivity, 1, **30–2**, 38
rights *see* ecological rights;
  environmental rights; human
  rights
Rio conference *see* United Nations
  Conference on Environment and
  Development

scientific networks
  forest policy, 66
  IPCC epistemic communities, 52–3,
    56, 59
social conflicts and sustainability,
  3–4, **32–4**, 41
'South' *see* 'North–South' differences
state *see* nation state
sustainability, 2–4, 8–9, **29–43**
  climate change management, 56
  economics, 31–2
  forest conservation, 6, 61
  human rights, 11
  justice, 16, 29–31
  labour, 34–7
  lifestyles, 37–9
  nation states, 4, 8
  politics, 39–42

sustainability – *continued*
  societies, 32–4
  TNCs, 8, 89, 96, 98–101

taxation, 15, 22, **186–7**, 189
technology transfers, 50–1, 188
TNC *see* transnational corporations
Trade Related Aspects of Intellectual
  Property Rights (TRIPs
  agreement), 91
transnational corporations (TNCs)
  codes of behaviour, 2, 8, **92–4**
  environmental accountability, 7–8,
    88–101
  environmental governance, **94–9**
  political institutions, 13, 167–8
  proposed convention, 8, **93**, 99–100
  regulation models, 8, 89, **92–4**
  regulation rationale, 88, **89–92**
truth, 108–9, 111, 113, 115, 161–2

United Nations
  climate change management, 5, 46
  code for corporate responsibility, 8,
    88–93, **94**, 99
  cosmopolitan democracy, 8, 98
  environmental refugees, 7
  forest policy, 65
  history and reform, 20–2, 25,
    167–8, 172, 174–5, 193–4, 205,
    224–6, 228–9, 233, 235
  human rights, 120–1
  juridical model, 214–15, 219
United Nations Centre on
  Transnational Corporations
  (UNCTC), 88–9, 94

United Nations Conference on
  Environment and Development
  (UNCED) ('Earth Summit'), 2,
    215
  ecological human rights, 121, 130
  forest policy, 61, **63–4**, 65, 68, 70–1
  humane governance, 20, 225,
    229–32
  TNCs, 13, 89, 91, 93–4
United Nations Development
  Program (UNDP), 224
United Nations Environment Program
  (UNEP), 45–7, 224

values *see* ethics

war, 17–18, 202, 207–10
  see also humanitarian interventions
world citizenship *see* planetary
  citizenship
Westphalian sovereignty, 17, 19, 67,
    234
women, as environmental refugees,
    73, 76, 79, 81, 83
World Business Council for
  Sustainable Development, 91, 94
world citizens *see* planetary
  citizenship
World Climate Research Programme
  (WCRP), 50
World Commission on Environment
  and Development (WCED), 29
World Environment Council / Forum
  (proposed), 9, 22
World Meteorological Organization
  (WMO), 45–7
World Weather Watch (WWW), 50